Endorsements

"The manuscript is well written and well argued, and surely it will be an important addition to the God-and-science literature." "A good and interesting book. Congratulations!"

—FRANCISCO J. AYALA, 2010 Templeton Prize winner and Donald Bren Professor of Biological Sciences, University of California, Irvine. Dr. Ayala is a noted author on the history of science and science and religion.

"The authors have an unusual approach to rank the probabilities of various arguments being convincing for the existence of God."

—OWEN GINGERICH, Harvard University, Emeritus Professor of Astronomy and History of Science.

"This book includes some of the hottest topics in science today, such as the relationship of the brain to spirituality, how inanimate matter became life, and the evolution of conscious, self-aware life." "The authors write in an engaging style that makes an entertaining read."

—GORDON LEIDNER, author of the book *Of God and Dice: Quotes from Eminent Scientists Supporting a Creator.*

"I think this book would be of interest to theologians, biblical scholars and to my colleagues in Southern Methodist University's Graduate Program in Religious Studies. It is written by two people eminently positioned to collaborate in this joint process from their areas of individual expertise." "This book is written with accessible depth, so that academics as well as general readers will benefit from it."

—Alyce, McKenzie, Le Van Professor of Preaching and Worship, Altshuler Distinguished Professor, Perkins School of Theology, Southern Methodist University. Dr. McKenzie is the author of the book *Hear and Be Wise: Becoming and Preacher and Teacher of Wisdom.*

"Your book is very well written. It covers the subject thoroughly and in language everyone can understand." "I can see how your book will be appreciated by thoughtful people who are looking for what is provided here."

—EMERY PERCELL, retired clergy—University of Chicago background in ethics, classics, religion, epistemology, and church history and forty seven years as a successful United Methodist Church pastor.

"Your assignment of confidence limits each of the evidences, which allows for building a cumulative case for God's existence." "I think your idea of the various aspects of 'tuning' is a novel one."

—RUSSELL STANNARD, a high energy nuclear physicist engaged in the study of the ultimate structure of matter and the properties of space and time. Dr. Stannard is the author of the book *Science and the Renewal of Belief.*

"A remarkably original and compelling book." "This book deals with the most important of all questions."

—RODNEY STARK, Distinguished Professor of Social Sciences and Co-director Institute for Studies of Religion, Baylor University. Dr. Stark has published 28 books including *America's Blessings: How Religion Benefits Everyone, Including Atheists.*

". . . an interesting and thought-provoking book." ". . . you reached your conclusions in an intelligent and fair-minded way."

—JIM STEIN, Professor of Mathematics at California State University, Long Beach and active in mathematics education for the State of California and the National Science Foundation. Dr. Stein is a nonmilitant atheist and a published author.

God Is Here to Stay

God Is Here to Stay

Science, Evolution, and Belief in God

Thomas R. McFaul

and

Al Brunsting

WIPF & STOCK · Eugene, Oregon

Wipf & Stock
An Imprint of Wipf and Stock Publishers
199 W. 8th Ave., Suite 3
Eugene, OR 97401

www.wipfandstock.com

ISBN 13: 978-1-62564-232-2

Manufactured in the U.S.A.

"I dedicate this book to my late parents, Carson and Dorothy McFaul, who taught me many life lessons as a growing child. They reminded me often of the importance of being honest, fair, and kind no matter where life might lead us. Not only did they speak with love and wisdom, they lived it as well; and they always took time to care for my brothers, sister, and me. I am grateful beyond words."

—Tom McFaul

"I dedicate this book to my late parents, Alice and Bernard R. Brunsting, who played a hugely positive role in my formative years. Mom encouraged me towards excellence in my goals and how to be compassionate with others through her example. Dad was an empathetic and effective career pastor, published author, and he always seemed to have time for me and the rest of his family. Mom and dad, with deep love this book is for you."

—Al Brunsting

Contents

Illustrations

Figures

Tables

Preface

The purpose of this book is to answer one question: what does the best scientific knowledge suggest regarding whether or not God exists? As will become clear, this question is not easy to answer for two main reasons. The first involves becoming aware of the enormous amount of scientific evidence that is available and of sorting through it. The second entails arranging the evidence in a way that makes one's position as persuasive as possible one way or the other.

Disagreements regarding whether or not God exists are not new. As is well known, they date back thousands of years in regions of the world as diverse as ancient Greece and India. Whatever position one takes regarding God's existence, one thing is certain: The evidence is not clear cut one way or the other. If it was, there would be no debate. Everyone would agree. Instead, dissent continues because individuals interpret the various pieces of evidence in different ways, and this leads them to opposite conclusions.

At the same time, despite the multiple interpretations and conclusions, all of the arguments both pro and con on the question of God's existence fall into two broad frameworks. The first is called naturalism, materialism, or atheism because it is confined to evidence drawn exclusively from the physical universe. It rests on the assumption that no hidden reality exists beyond the material forces of nature. For atheists the world as experienced through the five senses (seeing, hearing, tasting, touching, and smelling) is all there is. The cosmos is governed exclusively by cause and effect relationships that can be known through the rational-empirical methods of modern science. There is no need to presuppose the existence of God, gods, or any kind of Supreme Being that exists above, beyond, beneath, or within the material universe or that intervenes in it in any way through miracles or by suspending the laws of nature.

During the past decade many authors have become identified with an updated version of this ancient materialist tradition. They are called the New Atheists who continue the long term practice of critiquing traditional theistic views by arguing that modern science renders accounts of supernatural miracles and the belief in intercessory prayer unacceptable. Unlike past atheists, they are far more aggressive at promoting their non-theistic views through widespread public speaking, debate, and a steady stream of major publications.[1] In part, it is against the backdrop of this New Atheism that we have chosen to examine very carefully whether scientific knowledge increases or decreases our confidence that God exists. While we take the New Atheism seriously, one of our goals in writing this book is to make our own judgments after examining for ourselves the best scientific evidence available.

The second framework goes by various names such as theism, supernaturalism, among others. It differs from naturalism in that it posits the presence of a Supreme Power that transcends, created, and sustains the physical universe. From this basic starting point, the proponents of this second position can be found throughout the diverse cultures and religions of both the East and the West. While they share a common assumption regarding the idea of transcendence, they differ in their views of how this power created and becomes immanent within the natural cosmos and human experience.

As we progress through the following eight chapters, our aim is to draw on the best scientific evidence available to determine which of these two broad frameworks has greater credibility—atheism or theism. Our journey through the topics that appear in the following eight chapters starts with an Introduction that includes important background information and many of our assumptions about the nature of science, knowledge, and belief in God. Building on the foundation of chapter 1, we examine in chapters 2 through 7 the essential evidence that pertains to the central concern of this book: to decide whether it is more credible to believe that God exists than that God does not exist.

Chapter 2 looks at several areas of knowledge and the impact that modern science has had on each of them. The first topic covered in this chapter is communication and the last human health and longevity. This leads to chapter 3, which considers the connection between the universe

---

1. The New Atheism movement dates from the 2004 publication of Harris, *The End of Faith: Religion, Terror, and the Future of Reason.* This book prompted follow up writings by other adherents of the New Atheism. Among the best known are Dawkins, *The God Delusion;* Stenger, *God: the Failed Hypothesis—How Science Shows That God Does Not Exist;* Hitchens, *God Is Not Great: How Religion Poisons Everything;* Onfray, *Atheist Manifesto: the Case Against Christianity, Judaism, and Islam.*

and life. It describes how to interpret this relationship in terms of the Anthropic Principle. Chapter 4 explores the various scientific interpretations of how and when non-living matter that originally evolved on the Earth's surface made the jump—so to speak—to carbon life. What can we conclude about the process by which non-life became life?

The major issue presented in chapter 5 covers how the brain is linked to the experience of human spirituality. As we sort through some of the best research available on this intriguing topic, we will lay out what we can reasonably conclude about this connection. The next two chapters shift away from the realm of the physical and biological sciences and move into the arena of ethics and social experience. Chapter 6 investigates the matter of justice and how it manifests itself across the world's diverse communities. This leads to the larger issue in chapter 7 of whether or not a universal morality sits at the center of all societies despite dissimilarities that appear at the level of cultural norms and behavior. After examining the several topics that appear in chapters 1 through 7, we turn to chapter 8 where we pull together the collective evidence that we presented throughout the course of the book and draw our overall conclusion regarding the existence or not existence of God. We end the book with a brief Epilogue in which we identify issues that call for follow up discussion.

Before turning to chapter 1, we believe that it is helpful to provide some background information about ourselves as the coauthors of this book. We come from very different backgrounds—the social sciences, humanities, and physical sciences. McFaul is the social scientist and humanist who has spent most of his adult life in academic settings where he has dedicated over forty years to teaching and scholarship in the fields of sociology, philosophy, ethics, and comparative religions. He has written numerous articles on topics ranging from bioethics to the emerging global village. He has authored a three volume trilogy on the future of peace and justice, truth and freedom, and God, and a book that focuses on the formation of the moral imagination as a transformational process.

Al Brunsting is the physicist who has a track record of fifteen issued US patents, forty publications in refereed scientific and technical literature, editor of the Patents Review Board of Applied Optics, designer and developer of fiber optics products applied to telecommunications, and an international science award winner (Otto-Bayer Award) for substantial contributions in self-blood glucose monitoring.

For some, fruitful collaboration between a sociologist-humanist and a physicist with such divergent backgrounds might appear doomed from the start—like trying to mix oil and water. However, for the two authors who have shared many years of friendship, this is not the case. Despite their diverse mosaic of interests, a mutual focus fuels their curiosity—a deep desire to examine the complex connections between knowledge, science, and belief in God. This book is an outgrowth of their personal and professional passion to explore this three-way relationship and the hours of conversation it inspired.

Throughout the years that McFaul taught Comparative Religion on diverse campuses, he was privileged to work with many students whose questions and comments spoke wisdom beyond their ages. He learned early that some of the quietest students are often the most thoughtful. On one occasion, near the end of a semester after spending weeks examining the world's myriad spiritual traditions, a student who had not spoken a word all term raised her hand. After expressing her eye opening amazement at discovering for the first time the variety of beliefs that exist among the great faiths, she asked the question that sooner or later crosses the mind of every probing student: "Which one is right?"

It was an ideal teachable moment, and McFaul responded to her question with a question of his own. "Why is this important to you?" After decades of teaching he thought that he had heard it all, but her response caught him by surprise. It was unprecedented. Not only did it reveal the deepest desire of her heart but of humanity's heart as a whole. The room was hushed; and after a brief pause she looked up and spoke, "Because I don't want to make a sucker's bet." After years in the classroom, that single comment summarized one of the major motives driving McFaul's pursuit of the truth and underpinning his purpose for co-writing this book—to avoid making a sucker's bet at the deepest level of meaning in life—not only for himself but for others as well.

For Al Brunsting, it was not a late career classroom encounter with an inquisitive student that helped bring him to the doorstep of co-writing this book. Rather, his motives and purpose were shaped much earlier in life. At the impressionable age of thirteen, he and his family received word that his three year old brother Danny was diagnosed with leukemia. Despite follow up doctors' visits and intensive drug therapy, twelve months later Danny died, which Al witnessed. The emotional impact this had on Al's life cannot be overstated. He still remembers this event like it happened yesterday.

Given that Al's father was an esteemed minister in the Reformed Church of America and had dedicated his life to caring and spreading the word of God's love, Al began to wonder about the existential implications of his younger brother's death. Did Danny's life have any meaning? Was it just a random event? Did he pass into a better existence? Will we ever see him again? Is there a God? How do we know? What is the evidence? As a physicist, Al has spent his entire adult life searching for objective answers to these questions based on a scientific understanding of how the universe works. In the following chapters, as the coauthor of this book, Al shares some of what he has discovered.

Despite the dissimilarities that have shaped the backgrounds of the two authors, their interests as expressed in this book converge around the question of whether modern scientific knowledge reinforces or undermines belief in God. Anyone whose curiosity pulls them in the direction of searching for a single and undisputed response to this concern will quickly discover that no consensus exists. There is no one definitive answer—no unequivocal once and for all times "ah ha" moment—that either confirms or negates belief in God's existence. At the same time, a careful examination of the best scientific evidence available is an important place to embark on the journey that starts with chapter 1.

Acknowledgments

Numerous people have played important roles throughout the course of writing this book. Their thoughtful comments and recommendations have contributed in no small measure to its overall direction and development. We thank Jim Nelson, PhD, chemistry, and Robert T. Lehe, PhD, philosophy, who made many substantive contributions. Words cannot express our heartfelt appreciation. We are most grateful for the constructive suggestions that inspired us to sharpen our focus and improve the book's overall quality. Any errors or shortcomings that appear on the following pages fall strictly on our shoulders.

1

INTRODUCTION

Knowledge, Science, and Belief in God

As stated in the Preface, the primary purpose of this book is to explore the connection between knowledge, science, and belief in God. In order to examine the interaction effects that these three areas have on each other with as much clarity and depth as possible, we start by defining the major concepts and overall orientation that we will follow throughout the remaining chapters. Our central concern focuses on the question of whether or not the current state of knowledge based on scientific evidence points in the direction of believing or not believing in a purposeful God who created the universe. We begin our investigation by examining the difference between inductive and deductive reasoning.

Induction versus Deduction

Our methodology is based on the rational-empirical means of inquiry that emerged in conjunction with the rise of modern science during the past several centuries. At the height of the European Medieval Era during the thirteenth century CE, it was Roger Bacon (1214—1294) who first suggested that scientific certainty should be based on experimentation and the direct observation of nature rather than by appealing to sacred texts, ancient thinkers, or outside authorities.[1] Bacon advocated the now widely accepted

1. Dampier, *A History of Science and Its Relationship to Philosophy and Religion*, 90.

view that the search for scientific truth should proceed inductively from the "bottom up" starting with empirical investigation and not deductively from the "top down" based on longstanding religious or philosophical premises about the origin and operations of the natural order.

Although the vast majority of current scientists have adopted Bacon's approach, a word of caution is in order. We must be careful not to conclude that the inductive procedures of modern scientists exclude all forms of deductive reasoning. Induction and deduction are not mutually exclusive ways of thinking. Each includes aspects of the other. For example, except for pure mathematical logic, such as $2 \times 2 = 4$, deductive judgments often incorporate references to nature, such as the eastern religious belief that the souls of deceased persons reincarnate into new physical bodies.

Likewise, conclusions based on inductive investigations include assumptions about the operations of nature. There are many who presume the course of the cosmos is controlled exclusively by the laws of cause and effect or by randomness and not by a divine intelligence or intervention of any kind. Our approach assumes that the connection between deductive and inductive approaches to discovering the truth is subtle and complex.

Assumptions often guide perceptions, which can easily lead to circular reasoning. For example, believing that God exists predisposes one to "interpret" the "facts" to support this assumption. However, assuming the opposite leads one to interpret them differently and conclude that God does not exist. We are keenly aware of this tendency of the human mind to engage in circular reasoning. At the same time, our approach minimizes this risk even though it does not eliminate it entirely.

Evidence versus Proof

Our next step is to distinguish between evidence and proof. Evidence, as we understand this concept, is not identical to proof. Proof refers to certainty beyond doubt. Evidence refers to a recurring pattern of relationships that point in the direction of proof but that may fall short of certainty. Evidence includes signs, indications, and information that a given conclusion is valid or true although it might not be. The following example demonstrates this distinction.

In the novel, *To Kill a Mockingbird*,[2] Atticus Finch is a court appointed attorney for Tom Robinson, an African American man. The evidence that

2. Lee, *To Kill a Mockingbird*, 1960.

Finch presents in court indicates that Robinson is innocent of the accusation that he raped a young white woman. The racially biased jury ignores the evidence and convicts Tom Robinson. In this example, evidence implies four elements: 1) information is presented that is applicable to the subject (testimony), 2) a person or group of people (in this case the jury) interpret the information, 3) conclusions of that interpretation are formed by that person or group (the verdict), and 4) those conclusions are communicated to others (for example, the criminal justice system).

As is clear in this example, evidence is not identical to proof. However, using Euclidean Geometry, we can prove that the sum of three angles within any plane triangle is 180 degrees. In this case, proof uses a small set of starting points called postulates, assumed to be indisputable, and proceeds with logical and accepted steps to a conclusion. The conclusion—or proof—is independent of the reader. If one accepts the starting points and follows the accepted rules step by step, one must arrive at a conclusion or proof. There are no other alternatives, even considering the biases of the reader.

There are also other views of proof that are less restrictive than those that apply to mathematics and logic. They are based on rational-empirical evidence that is derived from scientific observations that scholars make in different fields of inquiry ranging from the physical and human sciences to the humanities. Conclusions based on empirical evidence are subject to biases, as in the case of Tom Robinson's conviction. As new evidence emerges, beliefs that were once thought to have been proven become subject to change unlike those of mathematics and logic.

We recognize that interpreting information (step two above) involves prioritizing in terms of importance. In *To Kill a Mockingbird*, the jury gave much more weight to the victim's testimony and little or no weight to Tom Robinson's. This is easily recognized as a selective bias or as it is often called "cherry picking" the evidence. The distribution of weight, or credibility, is clearly based on the jury's racial prejudices.

When we examine the knowledge that is related to the topics that we include in the remaining chapters of this book, we are mindful that our own selective biases could influence our choice of scientific evidence. At the same time, we are deeply committed to being as objective as possible. Our goal is to examine all the relevant and verifiable information and interpret it (step three above) as free from bias as possible.

By following this method, we believe that our weighing of evidence will be an accurate reflection of the real world, which we assume exists independently of our perception of it. Furthermore we assume that the real world contains elements that are knowable and verifiable by others as well as ourselves. To the extent that this is humanly possible, we are committed to scientific objectivity. If the jurors in *To Kill a Mockingbird* had recognized their biases, their verdict would certainly have been different and the outcome less tragic.

Before we conclude this section on the relationship between evidence and proof, we need to address one final point that applies to the process by which knowledge changes. As we will show in the remaining chapters, because of modern science our understanding of how the world works has undergone a dramatic transformation in the past 300 years. For example, as a result of the accumulation of scientific evidence, we believe that the Earth is not flat. While this cannot be proven beyond the shadow of doubt, the evidence that it is round is so overwhelming that it is not unreasonable to interpret the evidence as proof or as near as possible to proof. In addition, if a mathematical equation can be applied to a recurring phenomenon of nature, such as Newton's Universal Law of Gravitation,[3] then it is accurate to interpret this relationship as proven or as close as possible to being proven.

Evidence and Confidence

One of the best ways to envision the relationship between evidence and proof is to think of a continuum with disproof at the one end and proof at the other. See Figure 1.

Figure 1. The Evidence-Confidence scale and two extremes.

3. We recognize that Newton's Universal Law of Gravitation is superseded by Einstein's General Theory of Relativity.

Evidence can range along this continuum between these two extremes. The relative strength of evidence is directly related to the movement of the needle along the Confidence scale: the stronger the evidence, the greater the confidence.

The essential factor for determining any given confidence level is the reliability of evidence that is related to a specific scientific hypothesis that involves a predictable relationship between two or more elements. For example, if independent scientists observe a repeatable pattern of relationships that persists over time, then the evidence that they accumulate moves the needle toward the proof end of the continuum. As a result, the confidence level increases. However, if the accumulated evidence is inconsistent or not repeatable or contradictory, such as the causes of personality variations, the needle goes in the opposite direction toward hypothetical speculation or disproof. In turn, this leads to a decrease in confidence.

Scientific Method

Next, our understanding of evidence cuts across a broad range of areas that include science, knowledge, justice, and morality—all of which we describe in this book. We begin our discussion by focusing on the methods that scientists use to derive knowledge, which in turn affects our level of confidence. Our approach is eclectic.[4] That is to say, we accept that scientists use a range of methods that are appropriate to their separate fields of study. No single method applies to all areas. For example, the double-blind experimental laboratory methods that medical science uses to develop new pharmaceutical products differs from those that anthropologists employ when they do comparative field studies of diverse cultures. The sampling methodologies that sociologists utilize to determine political views within a large population differ from the modeling techniques that meteorologists apply to detect changing weather patterns. We also realize the limitations of the number and types of scientific cases we can present. We suggest that other cases can be treated in a similar manner to the ones we use here.

Regardless of their chosen method of investigation in the form of experimentation, comparison, observation, description, and/or modeling, all scientists follow the same systematic procedure. This involves several steps starting with the development of theories and hypotheses about empirical

4. Blakstad, "Research Methodology."

and measurable relationships within a given field of inquiry. This is followed by collecting and analyzing empirical data; interpreting results; sharing outcomes with colleagues through journals, conferences, and other forms of communication; and improving the techniques of investigation or experimentation when outcomes call for further exploration. In cases where disagreements exist—and they often do—research continues and hypotheses are updated with the goal of eventually building a scientific consensus based on the most convincing empirical evidence and predictability.

While this brief description might sound oversimplified, we believe that it correctly captures the modern scientific approach to the discovery and development of new and accurate knowledge in numerous fields of inquiry. In the following chapters, we build on this foundation and incorporate it into our discussion of the diverse topics that we cover.

Structuring or Tuning

In addition, the notion of structuring or tuning plays a central role in our understanding of evidence and the conclusions we draw from it. By structuring or tuning we mean that an animal, person, or object has an inherent characteristic that defines its essence. For example, an acorn is structured to become an oak tree and not an eagle. This is its nature. Dolphins are tuned to swim in water and not to slither on the beach. Humans are tuned to be biological beings and not rocks.

As will become clear in the chapters that follow, both the physical world and human experience are structured or tuned in specific ways that reveal the persistence of patterns or evidence over time. Once we have identified these patterns in chapters 2 through 7, based on the best scientific evidence available, we will discuss in chapter 8 how this knowledge affects belief or disbelief in the existence of God. In particular, we will address the question of whether the structures that appears throughout the physical universe and in human experience are the result of random forces or occur through some kind of intentional design.

Theological Ambiguity

We need to identify another important assumption that guides the writing of this book. We believe strongly that God's existence can be neither proven nor disproven beyond the shadow of doubt. We refer to this condition as theological ambiguity. Furthermore, we do not speculate one way or the other on if, when, or how this issue will be resolved in the future. This means that our objective is not to use knowledge to prove or disprove God's existence. In this sense, we do not have a hidden agenda. Instead, our goal is to establish confidence levels with regard to God's existence or nonexistence based on the best scientific evidence available.

In addition, we recognize that our effort to combine science and theology is not unique. We are acutely aware that we stand on the shoulders of many extraordinary thinkers and writers who have gone before us or still live among us. For centuries, scientists, philosophers, and theologians have debated the question of God's existence. In the next section of this chapter, we summarize the pros and cons of some of these main arguments.

Arguments for God's Existence

Over the centuries, proponents for God's existence have advanced many arguments for the existence of God, but four of them stand above the rest. They are called Cosmological, Teleological, Moral and Ontological.[5] Below we summarize both the strengths and weaknesses of each. As will become clear, there are many writers who claim that these arguments support belief in God, whereas others draw the opposite conclusion.

Cosmological

The Cosmological position derives from the Greek word *kosmos* or world. Both Aristotle in ancient Greece and Thomas Aquinas in the thirteen century CE championed it to make the case for God's existence. While there are variations on the theme, at its core it rests on the notions of causality and totality and is easy to grasp. The basic assumption that advocates of the Cosmological argument make is that nothing in the universe exists by itself. Everything is an effect that is brought about by a previous cause or causes. This first assumption leads to a second, which is that ultimately everything in the universe is linked to everything else even though the finite

5. While there are numerous arguments for and against God's existence, these four are the best known among the many. See Peterson, et al. *Philosophy of Religion*, 173–256. Also see Kreeft and Tacelli, "Twenty Arguments for the Existence of God;" and "Hundreds of Proofs of God's Existence."

human mind cannot observe or measure all the intricate causes and connections. Nothing stands outside of this complex web of cause and effect relationships.

At one level, the Cosmological argument for God's existence is based on common sense. As humans, we take for granted that our everyday experiences do not emerge randomly in a vacuum but instead are the result of causes that preceded them. Even though there are gaps in our knowledge of what causes what to happen when events that we do not understand occur, we assume connections exist because we could not function day to day without believing that consequences do not materialize out of thin air.

For example, we perceive through observation and measurement that the daily movement of the Moon around the Earth causes the tides to wax and wane. In addition to analyzing the objective forces of nature, we humans willfully make decisions that cause consequences for good or ill. The choice to stay in the Sun too long without protection against ultraviolet rays causes skin burn, which leads to the next decision to apply a healing ointment. Causes result in effects that become new causes that lead to new effects, and so on. Virtually all of the scientific explanations about how the world works, which we describe in later chapters, presuppose a cause and effect paradigm. Most of science could not function without it, and the scientific knowledge that we cite builds on it.[6]

It is from this starting point that proponents of the Cosmological argument take the next step to belief in God. It was Aristotle (and later Aquinas who borrowed his ideas) who first imagined that the chain or web of cause and effect relationships that we humans experience in the finite world could be regressed, that is, through the imagination be extended backwards in time logically although not necessarily temporally or empirically. Building on this premise, he wondered how the first moving object got going and whether the chain of moving objects could be regressed infinitely without some kind of logical starting point. Based on pure logic alone, he concluded that the first moving object had to be started by something that was not moving, which he called the Unmoved Mover. Even though both Aristotle and Aquinas did not speculate about an empirical starting point, both believed that logic dictates there must have been one. Aquinas identified Aristotle's Unmoved Mover as the infinite and eternal God who created the universe.

6. Exceptions to a cause and effect paradigm include quantum mechanics, climatology, and fluid turbulence.

Critics of the Cosmological argument point out that if it is possible to logically conceive of an eternal Uncaused Cause that caused the first object to move as the first cause in nature's chain of causes and effects, then it equally convincing to think of nature as have always existed and therefore without a first cause whether this be understood from the perspective of logic alone or as a temporal sequence. In addition, if randomness enters into the chain of effects, then God might not be required to explain the nature we observe. If God's existence were self-evident based on logically regressing cause and effect relationships back to the start of the universe, there would be no argument. However, we know this not to be the case. There is no universal consensus on the issue that it is necessary to logically think that there must be a first cause called God. Thus, the Cosmological argument, whether construed logically or empirically, does not lead definitively to proving or disproving the existence of God. Instead, it gives rise to contrary conclusions. Does the Teleological argument avoid this dilemma?

Teleological

Whereas the Cosmological position is based on the concept of cause and effect, the Teleological argument focuses on the ideas of purpose and design. The one side starts with the view that everything in life exists for a purpose, that is, it exists for a specific reason that involves moving toward an end of some kind. For example, the purpose of the acorn (its reason for existence) is to become an oak tree (its end). For the proponents of the Teleological position, the world is comprised of an integrated set of parts or objects of nature that in combination contribute to sustaining the universe as a whole. No component is without a place because it connects purposefully either directly or indirectly to all other components.

The design side of the Teleological perspective is closely coupled with the concept of purpose. At its core, the argument from design rests on the perception that the universe is so complex and intricate in its details that it could not have been produced by chance alone. Just as every part contributes to the whole, the entire cosmos provides ample evidence of design. The design comparison most often made is to that of a clock or watch which was created to keep accurate track of time. In order for the watch to function effectively, each gear or digital circuit must connect to all the others that comprise to total mechanism. In chapter 3, The Universe Is Structured for Conscious, Self-Aware Life, we will discuss the design issue in depth.

Using the image of a watch, it is easy to take the next step to belief in God. A watch does not come into existence by itself. Nor does it design itself. Rather, it is both produced and designed by a watchmaker. By analogy, the cosmos did not come into existence by itself. Like the intelligent maker of the watch, an intelligent God created a purposeful and intricately designed cosmos.

Opponents of the Teleological argument for the existence of God call this a weak analogy. They hold that the simple mechanics or digital circuitry in a watch cannot be compared with the extraordinary complexity of the universe. While the comparison is easy to grasp, it is not believable because it confuses, metaphorically speaking, cats with dogs. Since the analogy is faulty at this most fundamental level, critics claim that it is just as compelling to believe that the complex cosmos arose through a process of godless randomness as through the purposeful design of an intelligent Creator called God. Other critics challenge the assumption that every part of the universe is purposeful relative to the whole. For example, the death of a diseased child does not appear to contribute to any purpose associated with an assumed all loving and all powerful God, which we will discuss in greater detail below under the Ontological argument for God's existence.

Thus, we are left to conclude that the Teleological argument is no more effective at overcoming opposite outcomes similar to those that are connected to the Cosmological disputes that we described in the previous section. How does the Moral argument stack up against these two?

Moral

The Moral view differs from both the Cosmological and the Teleological arguments by centering on the person as a moral agent or cultures as a whole and not on the operations of nature. According the Moral perspective, all individuals regardless of the diverse societies and time periods into which they were born possess an innate Moral Law. It is the same for everyone everywhere and there are no exceptions. No one stands outside the framework of this Law.

One of the most often cited examples that proponents use to demonstrate the existence of this Law is that all cultures condemn murder, which we define as the intentional killing of an innocent person or persons with malice of forethought. The fact that murder occurs in all cultures does not nullify the Law. Rather, wherever and whenever it occurs, societies reinforce their universal opposition by condemning it. In like manner, all societies have sanctions against stealing, lying under oath in a court of law, along with many other values.

While the members of diverse cultures differ in how they interpret the Moral Law, nonetheless they all seek consistency in applying general principles to specific laws and behaviors despite social changes that occur over time. For example, the norm of not harming others by driving an automobile one hundred miles an hour into a crowd of people would not be relevant in pre-technical cultures where cars do not exit but would be in industrial societies where they do. Nor does removing life sustaining medical equipment from permanently comatose patients apply to societies without respirators but only to those that have them. The specific application of universal moral standards must fit the level of evolution in any given culture. Societies adjust to social change by applying the general Moral Law in novel and consistent ways that match their specific level of scientific and technological development.

The case for the existence of God in the Moral argument extends from the perception that all societies show evidence of a shared set of ethical standards. The reason the Moral Law appears everywhere is that God created it as an inherent part of our human nature. No matter where or when people are born, this Law has always been and still is structured into every person's moral conscience. Why should this be so? The answer is that God put it there as many authors including C. S. Lewis have argued.[7] In chapter 7, "Humanity Is Structured for Universal Morality," we will revisit the issue of universal morality and examine it extensively.

While for many this line of reasoning sounds convincing, in parallel to the Cosmological and Teleological debates not everyone agrees with the pro-God interpretation of the Moral argument. This is revealed clearly through the voices of critics who express an opposing opinion. The most outspoken opponents are relativists who claim that the abundance of evidence leads to the conclusion that a universal Moral Law does not exist. They hold that the contrary is true as witnessed by the diversity of values that prevail from society to society and from individual to individual.

According to the cultural relativist viewpoint, the demands of survival impose on every society the imperative to create shared values that will insure continued existence. If they do not succeed, they perish. In the process, each must develop its own sets of shared standards that result in

---

7. Lewis, *Mere Christianity*.

group support and solidarity. Examples are plentiful. Eskimos in the Arctic tundra, tribal groups in the Amazon rain forest, and Bedouin clans in the Arabian desert all differ widely in how they dress, what they eat, their concepts of inheritance, gender relationships, punishments for norm violators, religious ideas, and so on.[8] Another example drives home this point. The collective values of the Amish in the state of Pennsylvania deviate substantially from those of the individualistic, high technology American culture that surrounds them. In addition to entire cultures, individuals within the same culture, as well as across cultures, can and do diverge widely in their value preferences.

In the eyes of cultural relativists, these examples and many others that could be cited lead to two overarching conclusions. First, the existence of cross-cultural value dissimilarities demonstrates that a universal Moral Law does not exist, and second, there is no need to invoke the concept of God to explain the origin of any values. A naturalistic explanation is all that is necessary. As disparate groups confront unique survival demands in dissimilar geographical and ecological settings, they develop dissimilar norms. A God centered approach is not required.

How do we assess the significance of this approach? Like the Cosmological and Teleological disputes, the Moral argument does not provide us with airtight proof one way or the other for the existence of God. This leaves us with only one remaining argument: Ontological.

Ontological

The Ontological argument was developed by the theologian Anselm of Canterbury (1033—1109). Among the four views that are discussed in this chapter, the Ontological argument is the only *a priori* argument. The other three are *a posteriori*. Anselm's position starts with the role of linguistic logic in relationship to the idea of God. Within the structure of any given language, it is possible to think of something beyond which nothing greater can be conceived. This is the essential meaning of the word God. If something greater can be conceived, then that which was originally believed to be the greatest is not God. Whatever specific term any given language uses, the idea of God holds a unique place within all languages because it can be used only in one way. By definition, the concept of God refers to an ultimate reality beyond which nothing greater exists.

8. Durkheim, *The Elementary Forms of the Religious Life*.

The next step in Anselm's deductive method involves assigning to the idea of God two specific attributes. The first is perfection. The phase that most aptly summarizes Anselm's argument is that God is the sum of all perfections. God is all loving, just, knowing, merciful, forgiving, and so on. If any being, according to Anselm, possesses any of these characteristics in greater abundance than God, then the original idea of God is erroneous and must be replaced by a concept of God that is superior to it and all other lesser beings.

The second attribute that Anselm assigns to God is existence. God must be more than an idea. What exists as a subjective concept in the mind must also have objective status outside of the mind. The reason the human mind can conceive of something beyond which nothing greater exists is because in reality nothing greater does exist. Of necessity, Anselm contends that it is only logical to conclude that existence and perfection are inseparable attributes of God. God's essence as the perfect being is not believable if no such being exist. Nor is it credible to accept the existence of a being beyond which nothing greater exists without assuming that such an objectively existing being is perfect.

While it might appear that Anselm's logical argument is an irrefutable proof for God's existence, not everyone agrees. From the time he first proposed it, the Ontological argument has met with a steady stream of opposition. While critics have expressed genuine appreciation of Anselm's attempt to prove that God exists through deductive reasoning alone, they are not persuaded by his position—for the following reason.

One of Anselm's contemporaries Gaunilo pointed out that just because one could conceive of a perfect island does not mean that such an island exists even if it is superior to all the other islands one could envision. In addition to Gaunilo, others have criticized Anselm's contention by referring to images of make-believe beings that do not exist. As they have pointed out, one of the fanciful images that many creative authors have included in their fictional writings is that of the one horned unicorn. While such a creature can be visualized in the mind, and might even exist in reality, no one has ever seen one. In short, although one can conjure up subjective mental ideas of something does not mean that this something exists outside the mind.

The pros and cons of the Ontological argument for the existence of God have been debated for centuries. Those who favor it claim that if it is possible to think of something beyond which nothing greater can be conceived, then in the world of possibilities God exists. Those who oppose the ontological argument hold that thinking of something beyond which nothing greater can be conceived does not mean that this something actually does exist. Thus, despite Anselm's Ontological argument that God exists, potentiality does not imply actuality. Or said succinctly, thinking does not make it so, *a priori* logic to the contrary notwithstanding.[9]

In addition, opponents of the Ontological argument for God's existence, as well as the Moral argument, point out that the pervasive presence of evil in the world negates one of Anselm's cornerstone contentions that God must embody the sum of all perfections. They question why an all loving and all powerful God would permit or even minimally tolerate cruelty, brutality, or unkind actions such as Al Brunsting's personal experience of seeing his younger brother die of leukemia or mass killings by persons suffering from inborn and severe psychotic disorders.

On one hand, if God is all loving and evil exists, then God cannot be all powerful. On the other, if God is all powerful and evil exists, then God is not all loving. God is either too weak to stop evil or less than all loving by allowing it to continue. Opponents of Anselm's view also hold that the argument that God gave us free will to choose between good and evil does not resolve this dilemma, especially in light of the enormity of such evils like genocide as exemplified in the Nazi Holocaust against Jews or in the slaughter of innocent children. Thus, from their perspective only one conclusion can be drawn from this irreconcilable conundrum: God according to Anselm does not exist.

Based on our discussion of the above four arguments for God's existence, what can we conclude? First of all, we are left with one overarching impression. Whether we focus on the Cosmological, Teleological, Moral, or Ontological argument or some combination of them, it is clear that for every "yes, God exists" there is a "no, God does not exist." This observation begs the question. Why is it not possible using either inductive or deductive reasoning to prove or disprove God's existence? From our perspective, there is only one answer to this question. As we stated in an earlier section of this chapter, the universe is theologically ambiguous.

If history is any guide, the outcome of the four classic arguments is a stalemate. None of these four arguments is widely accepted. Thus, we believe that the alternative of establishing confidence levels based on scientific

9. Kant, *Critique of Pure Reason*.

evidence is a superior approach to trying to either prove or disprove God's existence.

Creationism-Evolution and Theism-Naturalism

The dispute over God's existence does not end with the ancient arguments. There are modern manifestations of it as well. No discussion of the relationship of knowledge and belief or disbelief in God would be complete without a discussion of the impact that modern science has had on the interpretation and reinterpretation of pre-scientific scriptural texts. It would not be far from the mark for us to assert that more than any other factor during the past three hundred years, the findings of modern science stand head and shoulders above all other factors in challenging some of the most sacred beliefs that have held sway for centuries. Responses from the leaders and laity of the world's diverse spiritual communities have ranged along a continuum from complete acceptance to total rejection.

At one level, the story line is complex because reactions to the rise of modern science have varied so widely among people of different as well as similar faiths. At the same time, for some scholars this complexity can be divided into four distinct response types: science and religion are in conflict, they function independently of each other because science deals with knowledge and religion with faith, science and religion are in dialogue over different ways of knowing, and finally, their different points of views can be integrated. Others view the relationship between science and religion in terms of three distinct possibilities: complete acceptance of scientific findings by religious communities, partial acceptance in selected areas, and out-and-out rejection.[10]

The starting point for sorting out where to place different groups according to their reactions dates back to the rise of modern science itself and the role that religion played in stimulating its growth and development. Like any other historical movement, modern science did not emerge in a social vacuum. Rather it was nurtured within a cultural context that fostered its earliest impulses and promoted its continual expansion.

Anyone who examines how the relationship between science and religion developed during the past four centuries will discover quickly that the vast majority of historians of science are in agreement that the roots

10. Barbour, *Religion and Science: Historical and Contemporary Issues.* Barbour, *When Science Meets Religion.* Clayton, *The Oxford Handbook of Religion and Science.*

of modern science sprang from the soil of Western Christianity, especially in its Protestant form, coupled with influences from Islam and Greek rationalism. The discoveries of scientists and mathematicians like Copernicus, Galileo, Kepler, Newton, and Boyle, who contributed to our modern understanding of the cosmos are well known. What is less well known is that they and others believed that the laws of nature were created by an intelligent designer. Even Albert Einstein who did not believe in a personal God once commented that when it comes to the physical universe, "There are not laws without a lawgiver."[11]

Given that 1) Western Christianity kindled in no small way the advent and growth of science and 2) that many of the most important modern scientists found little or no incompatibility between their discoveries and belief in God, we are left to question how much of the current hostility that exists between these two spheres arose. In no small measure, the answer to this question can be found in the reactions of conservative Christians who believed that Darwin's theory of evolution and survival of the fittest through natural selection threatened to drive some of their most sacred beliefs into extinction.

It is ironic that just as Christianity inspired the emergence of modern science, a segment of conservative Christianity, especially those who have come to be called fundamentalists, rejected many of its findings. In addition, many conservative leaders and laity stood against the German theologians whose application of the methods of modern science to biblical studies called Higher Criticism led to rejecting the Bible's accounts of divine miracles and supernatural intervention or suspension of the laws of nature.

At this point, it is necessary to define how we are using the word fundamentalism in order to avoid later confusion. The dominant doctrine that emerged in the early church and that the Council of Nicea endorsed in 325 CE became known as Orthodoxy. As modern science emerged, the Christians who sought to reconcile its findings with their faith were labeled liberals or modernists whose ideas clashed with many traditional Christian tenets. Prior to this time, virtually all Christians accepted the traditional orthodox position. As the trend toward liberalization began to expand throughout Christianity, by comparison orthodox believers became the church's conservatives.

11. Hermanns, *Einstein and the Poet: in Search of the Cosmic Man*, 60.

The movement by Christians who reacted against this modernizing tendency because they perceived that it threatened the core beliefs of their faith came to be called fundamentalism, which can be viewed as the most conservative form of Orthodox Christianity. In response to liberalism, this group articulated a series of beliefs they called the fundamentals. As the modern evangelical wing of Christianity developed in the late nineteenth and throughout the twentieth century, its followers embraced these fundamentals in one form or another as the theological foundation of their faith.

For the purpose of this chapter, the story of how fundamentalist doctrines developed merits describing. In 1910, the conservative Christian brothers Milton and Lyman Stewart embarked on a five year program of soliciting comments from evangelical pastors, missionaries, theology professors and students, and diverse religious workers on the true nature of Christian belief. The outcome of this five year venture was the publication of a twelve volume collection of writings entitled *The Fundamentals: a Testimony to the Truth*. This multi-volume study consisted of a compilation of separate pamphlets that were written by numerous authors from 1910 to 1915 and that reflected the fundamental beliefs of a wide spectrum of conservative Christians.

Although *The Fundamentals* aimed much of its criticism at liberals, its main focus was on specifying the irrefutable core of truths without which the Christian faith would not be able to preserve its historical identity. This core consists of 1) the inerrancy of the Bible, 2) the literal nature of biblical accounts including Jesus' miracles and the *Genesis* account of creation, 3) the virgin birth of Jesus, 4) the bodily resurrection and return of Christ on Judgment Day, and 5) the substitutionary atonement of Christ on the Cross. The belief by some conservative Christians called dispensationalists and millennialists that Christ would soon return to judge the world was not shared by all fundamentalists. Despite this difference, *The Fundamentals* gave the emerging evangelical movement a shared doctrinal foundation.[12]

It was the second of the five fundamentalist beliefs that pitted evangelical against liberal Christians who accepted Darwin's theory of evolution and eventually the Big Bang theory of how the universe began in a massive explosion of energy 13.7 billion years ago. At this point a word of clarification is in order. The word evolution carries multiple meanings. It can refer to the development of the physical universe since creation, to Darwin's ideas, to the life cycle stages of various species, or any number

12. Ruthvan, *Fundamentalism: the Search for Meaning*, 10–26.

of other areas. In the case of Christian fundamentalists, the conflict over evolution turned on their rejection of the origin and age of the cosmos and of the Darwinian concept that life on Earth evolved through a process of natural selection.

As the conflict between Christian fundamentalists and liberals over evolution began to boil over during the early twentieth century, it eventually erupted into the well-known 1925 Scopes Monkey Trial in the small town of Dayton, Tennessee. As a high school teacher, John Scopes was accused of teaching the Darwinian theory of evolution contrary to the state's Butler Act that prohibited it. The trial, which drew national attention, was staged deliberately to draw attention to the modernist versus fundamentalist conflict between the biblical account of creation and evolution.

Two of America's most famous lawyers championed the opposing sides of the dispute. Colorful Clarence Darrow defended Scopes' position on evolution, and William Jennings Bryan, three time Democratic presidential candidate, supported biblical literalism. Even though Scopes was found guilty and fined one hundred dollars, his verdict was later overturned on a technicality. Despite the temporary sensationalism that the trial spawned throughout America and the media, it is the legacy of the trial that stands out above all else. It gave national exposure to the gap that was growing among Christians in particular and American society in general over the nature of truth and how to discover it: through biblical literalism or modern science.

The early success that literalists experienced in the wake of the Scopes trial gradually waned as the theory of evolution gained ground throughout America. The one area where this shift is most noticeable is in the writing of high school biology textbooks. During the late 1920s and early 1930s, few of these texts included reference to evolution to explain creation and life on Earth. Over time, however, this changed despite the efforts of evangelical Christians to reverse the tide. Eventually, the public began to demand that science textbooks be written by scientists and not educators who lacked in depth knowledge or promoted a religious agenda.

After the Soviet Union launched Sputnik into space in 1957, the American government passed the National Defense Education Act in 1958 so that the United States would not fall behind the Russians in science. This action led to the development of textbooks that emphasized the importance of evolution as the unifying theory in biology. Attempts by evangelicals to re-insert biblical creationism into textbooks as an alternative scientific theory to evolution met repeatedly with rejection.

All in all, trends in rising support for improving public education, Supreme Court decisions creating stricter boundaries between church and state, and increasing education levels throughout society served to boost support for evolution at the expense of biblically based accounts for creation and the development of life on Earth. While little known, the 1967 repeal of Tennessee's Butler Act that conservative Christians used to bring Scopes to trial in 1925 symbolizes as much as any other event the declining public support for biblical literalism.

However, the evolution vs. creationism story does not end here as conservatives pushed hard to develop a Creation Science that would be able to compete with the theory of evolution for scientific credibility. During the 1960s, the evangelicals' strategy shifted from opposition to Darwinian views to the assertion that the scientific evidence supports the literal interpretation of the Bible. However, despite these efforts, the American public in general continues to trust the findings of science as reflected in persistent support for the theory of evolution even though evangelical literalists continue their struggle to discredit it.

Thus, based on the above discussion, we can say with confidence that trends over the past fifty years have shifted in the direction of growing support for scientific evolution and away from biblical creationism. At the same time, however, one of the unfortunate byproducts that *The Fundamentals* and the evolution versus literalism dispute it helped create was to tie belief in God to biblical inerrancy. For many thoughtful persons, this implies that there are only two alternatives: either to accept the findings of modern science and deny the existence of God or to believe in God and reject the discoveries of science. This is the legacy of Christian fundamentalism.

At this point, a word of caution is in order. We must be very careful not to reduce the connection between science, evolution, and belief in God to a battle between pre-scientific biblical literalism and scientific modernism. Many conservative and liberal Christians accepted Darwin's theory or some version of it. What was a stake was something deeper: whether or a theory of evolution in any form would eventually lead to a purely naturalistic interpretation of the origin and development of the universe.

This is a complex issue that operates at two levels, which need to be kept separate. At one level, Christian fundamentalists disagreed with both Christian and non-Christian liberals over the issue of how universe came into being and evolved. It is at this level that the fundamentalists' legacy that to accept God meant to reject science and vice versa emerged. At the other level, all Christian theists felt threatened by the naturalistic implications of

Darwin's ideas, no matter where they stood on the theological spectrum or what they thought of evolution. What united both fundamentalists and non-fundamentalists was their perception that Darwin's views would undermine a theistic view of creation, however this occurred, in favor of a naturalistic one.

As a result, in support of theism, they found a common voice by responding in two main ways. The first involves the development of a Creation Science that was designed to defend biblical literalism. The second, which incorporates evolution in some form, entails a broad critique of the inherent limitations of science itself. In combination, they gave rise to two perspectives that are often invoked in order to counter a purely naturalistic interpretation of the cosmos. They are called God of the Gaps and Intelligent Design, which we discuss in the next two sections of this chapter.

God of the Gaps

The God of the Gaps idea is easy to grasp. When modern science cannot explain what causes what to happen in a sequence of cause and effect relationships, a knowledge gap exists. Or stated differently, there is a missing link in the explanation (or hypothesis). When this occurs, the supporters of the God of the Gaps view fill in a god where there are holes in our understanding.[13] It is an argument that is based on what we do not know rather than what we do know. It is an argument that Creation Science used to undercut the credibility of the theory of evolution and that theists who embrace some form of evolutionary science employ to argue against the philosophy of naturalism.

We reject this approach because it is not based on evidence but instead on the lack of it. In addition, to resort to a "God causes it" argument as a way to explain what we do not know or think we know but really do not because we lack accurate knowledge is a set ourselves up for failure. Since the start of the nineteenth century, in one field after another, scientific discoveries and explanations have replaced the god hypothesis as a way of understanding the cause and effect relationships that occur in natural and social settings.

Examples are many, but three will suffice to illustrate this point. Based on scientific evidence, we know that the common cold is not caused by

13. Collins, *The Language of God*, 93, 95, 193–95, 204.

divine punishment for sin, as was once believed, but by germs. We also know that a god does not cause the Sun to rise or set. In fact, the Sun does not rise or set at all. We only perceive it this way. Instead, it appears and disappears as a result of the Earth's daily 360 degrees rotation during its 365 day journey around the Sun. According to modern astronomers, the universe did not come into existence 6,000–10,000 years ago through divine command but rather because of a sudden and spectacular Inflation called the Big Bang explosion that dates back 13.7 billion years.

At this point a word of caution is in order. While we refuse to accept a God of the Gaps hypothesis as a substitute explanation for lack of scientific understanding, we neither reject belief in the existence of God nor seek to push God out of the picture through some form of atheistic or naturalistic reductionism or that one day science will explain everything in terms of natural causality. We recognize that there are still many events and relationships in the physical universe that cannot be explained through the discoveries of modern science.

At the same time, we do not take the position that the advancements that science has made to date imply that one day science will explain everything that goes on in the universe through purely naturalistic explanations. Such an assumption falls outside the methodology of science and into the metaphysics of naturalism. Our goal is not to debate this issue but to examine scientific evidence in various areas of inquiry as described in chapters 2–7 in order to establish confidence levels regarding belief or disbelief in God's existence, which we do in chapter 8.

Intelligent Design

The second approach is called Intelligent Design (ID), which is closely coupled with the God of the Gaps perspective. However, unlike a God of the Gaps explanation that is directed at causality gaps for which science cannot give an account or has yet to provide one, the ID perspective that involves multiple interpretations and applications rests on the assumption that at their deepest level the complexities of nature alone offer sufficient evidence for negating the belief in evolution or naturalism. Creationists and other theists often refer to the human eye and the nano-structures of life to support their view that the cosmos cannot be explained through evolution or naturalism because of its intricacies. In a nutshell, many theists and creationists in particular hold that evolution "is fundamentally flawed, since it cannot account for the intricate complexity of nature."[14]

Like the God of the Gaps perspective, we reject the ID approach because it assumes that modern science is permanently incapable of explaining the causality links that appear in nature not only because of inexplicable gaps but because there are some areas of nature that remain inaccessible to modern scientific methodologies. In this sense, the ID argument parallels the God in the Gaps approach and that together they provide biblical literalists as well as liberal theological evolutionists with a twofold rationale for rejecting either the Darwinian view of evolution in the narrow sense or a purely naturalistic explanation for the causal relationships that exist in the universe in the broader sense.

We believe that both of these approaches exist to save a theistic interpretation of the universe by building an impenetrable fence beyond which science cannot go. In essence, in order to show that despite the extraordinary progress that science has made in explaining the evolution of the universe, there are gaps and complexities that leave space for arguing for theism. Given that we believe that we cannot use the findings of modern science to prove or disprove God's existence, for us the only question is whether modern scientific knowledge increases or decreases our confidence in the existence of God. We are not looking for ways to block science by looking for gaps or complexities that are inherently beyond the reach of science. If scientific evidence decreases our confidence, then science reinforces an atheistic interpretation of nature and our human understanding of it. On the other hand, if the best scientific evidence available increases our confidence that God exists, then we are left to ponder how we might interpret this from a theistic perspective.

We are now ready for the next chapter where we focus on the role that knowledge has played throughout the long haul of human history.

14. Collins, *The Language of God*, 184.

2

Humanity Is Structured for Knowledge

Introduction

We begin this chapter by noting that humanity is different from all other non-human species in one important regard. We are purposeful knowledge seekers who labor to learn as much as possible about how the world operates by adjusting to or changing it in order to function as effectively as possible within it. We also desire to know how the world works as an end in itself with little or no connections with practical applications. This means that as modern humans, *Homo-sapiens*, we endeavor to survive, to thrive, and to understand. We do this in a way that is different from all other types of animals. It is through evolution that we have become structured or tuned for knowledge.

No doubt, as in the case of humanity, the capacity to adapt and endure generation after generation has been demonstrated repeatedly by the Earth's enormous variety of animal and nonhuman groups. At the same time, because of changes that developed in the brain, which we will describe in greater detail in chapter 4, human evolution followed an alternative pathway that is distinct from all other species. Whereas nonhuman adaptations are commonly linked to genetic programming or inborn biological instincts, human advancement mostly depends on intelligence and the capacity of the human brain to operate through memory, critical thinking, and high levels of linguistic abstractions called language. This allows human adaptations to occur much faster than reliance on only biological (morphological) changes.

Before we proceed with a detailed description of how humans create and use knowledge and in order to avoid later confusion, it is important to differentiate nonhuman consciousness from human self-consciousness. Consciousness in general refers to an organism's or creature's awareness of its circumstance along with the ability to adjust to it in order to assure intergenerational survival. All nonhuman species have nervous systems, ranging from simple to complex, through which they possess awareness of their external surroundings. Birds fly through the air, fish swim in water, and ants walk on the Earth. These and all other animal species are aware or conscious of the environments in which they operate and in which they survive through instinctive evolutionary adaptation.

The same can be said of us, *Homo-sapiens*—and more. Whereas modern humans share consciousness with all other nonhuman species that possess nervous systems, only humans possess self-consciousness. The difference, of course, rests in the word "self." Only people are aware that their actions and states of being originate within the self. Individuals everywhere and in all cultures reflect on questions about their origins and destiny. While all creatures die, only humans contemplate their purpose in life and whether there exists some form of life after death. Small worms and large whales do not. Also, only humans uniquely desire to know when and how the universe came into existence and where it is headed.

In addition to the distinction between consciousness and self-consciousness, it is necessary to define what we mean by knowledge. As stated, as nonhumans evolved, from aardvarks to zebras, nature coupled their consciousness of their surroundings with innate instincts that insured their survival. Bees know instinctively from birth how to behave in the hive. While humans share many of the same biological needs that other species possess, such as the need to satisfy hunger, thirst, reproduction, among others, human survival does not rest mainly on inborn instincts.

Instead, modern humans live primarily in culture, which mainly depends on the capacity for memory made possible through the development of the newer and higher part of the brain called the cerebral cortex. This in turn has given rise to the ability to intentionally create and manipulate mental sounds and symbols called language. Humans convert experience into knowledge that is stored linguistically in memory, is available for recall, and is communicated through sophisticated symbolic systems. While many nonhuman animals and primates demonstrate competence in communicating through various sounds and signs, complex patterns of human interaction throughout the world's diverse cultures would not be possible without more advanced cognitive capacities based on the use of abstract symbols.

The development of language enabled humans to deal more effectively with tasks such as migration around the world, hunting in groups, development of agriculture, living in towns, and the development of religion. These types of tasks helped to efficiently form the basis of civilization.

After language other types of communications that are associated with the emergence of civilization developed. Examples include art forms, arithmetic, algebra, geometry, and logic. These types of communication stimulated knowledge, which is humanity's storehouse of understanding of how the world operates. From the moment *Homo-sapiens* emerged on Earth about 200,000 years ago, each generation has passed on and continues to pass on to the next the totality of accumulated knowledge, which has grown with each subsequent generation. It would not be an exaggeration to say that the perpetuation of knowledge throughout the long haul of human history, based on language and memory, is probably the single most important factor in sustaining cultural development.[1]

If this progression of knowledge were to be interrupted for even one generation, human evolution on Earth, as we know it, would cease to exist. When we grasp that all nonhuman animal species survive by instinct and only humans do so by living in cultures that are rooted in the myriad symbolic and linguistic patterns we call language, we understand why preserving knowledge in the present and transmitting it to future generations is the one indispensable prerequisite that has insured the continuity of human communities throughout history.

At the same time, humans do not strive merely to survive but to enhance our standard of living. This is because humanity occupies a unique niche in the ecology of the Earth's evolution that is based on intentional learning. From the start, humans have created, stored, and passed on knowledge by observing, investigating, and drawing conclusions about their physical and social surroundings. As subsequent generations everywhere moved forward through time and spread out across continents, this cognitive heritage provided them with an ability not only to adjust to their

---

1. van Doren, *A History of Knowledge*. This volume covers the great theories and discoveries of humanity from the start of civilization to the present time.

surroundings but to manipulate and change them as well in order to enhance the conditions of life on Earth. Simply stated, for humans surviving
depends on expanding the pool of self-conscious knowledge, and thriving
results from knowing how to use it to respond successfully to the challenge
of survival.

However, as history has demonstrated repeatedly, it is not in our nature to sit on the status quo, so to speak, and merely submit passively to
things as they are or appear to be. Instead, we examine our surroundings
to understand what is going on and how things work. We define as merely
temporary the problems and obstacles that nature puts in front of us. We
gain novel understandings based on what we observe and learn. Then, because we are structured for knowledge, we apply this new capacity to push
ever outward the boundaries that block our purposes in order to achieve
them at progressively higher levels.[2] Just as we are inseparable from nature,
overcoming the restrictions of nature is a major driving force in our human
nature.

In the following sections of this chapter, we will describe some of
the most significant ways in which humans have applied knowledge to
overcome a number of natural barriers—and as a result have advanced to
higher levels of accomplishment. This in turn will prepare us for the topics
we take up in the remaining chapters of this book. In this chapter, we begin
with communication and end with human health and longevity.

Communication

At the most basic level, to communicate means to exchange information.
This occurs among all animal species, both human and nonhuman. At the
same time, when human communication occurs, it typically involves sharing information through the use of language. While humans and nonhumans communicate in other ways, such as through hand gestures, facial
expressions, body poses, and others, only humans communicate through
sophisticated pictorial or alphabetical languages that involve grammar,
syntax, and thought constructions in the form of sentences, paragraphs,
and extended writings. This unique evolutionary adaptation serves as
the basis for all other types of human transformation. When humanity's

2. Nisbet, *History of the Idea of Progress,* describes human progress in terms of some
of the most important cultural, social, scientific, and technological achievements from
antiquity to the modern era.

collective pool of knowledge is stored in memory or written down, it can be communicated linguistically to future generations who are able to use it to foster further human developments.

The story of human communication dates back for at least 150,000 years and starts with the early oral utterances and wall paintings of primordial tribes of *Homo-sapiens*. Prior to the conversion of vocal sounds into written symbols, the scope of any group's span of communication was restricted by the physical space it occupied or the new territory to which it migrated. Given the limitations of time and space that nature imposed on ancient cultures as they spread out around the world prior to the development of writing, small and isolated communities developed hundreds of spoken languages. When this ancient oral and local pattern is compared to the complex web of instantaneous global networks that exist in today's world, the transformation of humanity's capacity to expand the range and speed of interaction by developing innovative communication systems is nothing short of astounding.[3]

We will describe this amazing journey in two stages: first writing and then electronics. The first stage involves the evolution of writing and its related forms of expression. Between 3500 to 2900 BCE, the three ancient cultures of Phoenicia, Sumer, and Egypt took the first steps in transforming vocal sounds into abstract visual images we call language. The Phoenicians developed an alphabet, the Sumerians clay tablet pictographs called cuneiform writing, and the Egyptians created hieroglyphics they etched onto stoned surfaces. From 1775 to 1400 BCE, early history Chinese and Greeks also added to this list with writing innovations of their own. The impact that learning to write has had on human progress cannot be overstated. One source affirms, "With the invention of writing, the possibilities for human advancement increased dramatically. The written word is arguably the most important invention in the history of knowledge."[4]

In developing ancient cultures a need developed for a reliable means for transmitting and archiving information such as financial accounts, historical records, military conquests, and religious texts. About 3000 BCE, the complexity of trade, legal codes, religious oral traditions, and administration in Mesopotamia quickly expanded beyond the capacity for human

---

3. Poe, *A History of Communications,* explains the origins and effect of different forms of communication, such as speech, writing, print, electronic devices, and the Internet, throughout the long run of human history.

4. Goddard, *Concise History of Science & Invention: an Illustrated Time Line,* 15.

memory. Writing became a more trustworthy method of recording and presenting transactions in a permanent form.

Once the ability to translate sounds into durable inscriptions was set in motion, the rest of the story of communicating through the written word entails making further refinements in how to record and convey knowledge more efficiently and effectively. Over time, this led to the expansion of Eastern pictographic patterns and the development of Western alphabetical systems that contain a fixed number of reusable letters for fashioning a continuous flow of new words that in combination lead progressively to the creation of new knowledge.

Equally important with the emergence of writing has been the development of methods by which it is produced, preserved, and passed on to others. It is because of innovations in the technology of writing that the written word maintains its status as being the most important invention in the history of knowledge. Examples are numerous. The oldest record of writing in China dates from 1400 BCE and involved using bones as the medium for recording messages. Writers elsewhere wrote on clay tablets or engraved their thoughts and images on stone. Around 550 BCE, papyrus and parchments made of dried reeds served writers from bygone early African and European cultures as useful writing surfaces. As writing proliferated in the ancient world before the Common Era, the Syrians created the first encyclopedia and the Greeks started the first known library.

It was only a matter of time before someone created a more durable and portable inscription method. In the year 105 CE, Tsai Lun in China invented paper as we know it. The worldwide diffusion of this one innovation led eventually to the proliferation of an endless stream of bound books, pamphlets, magazines, newspapers, and a plethora of printed materials ranging from comics to catalogs. The demand among the masses for written materials in formats of all kinds stimulated the next and most advanced form of writing technology: the printing press.

The earliest Chinese technique of compressing wooden blocks onto paper inspired the invention of movable type. Building on earlier forerunner clay based printing machines in 1455 CE Johannes Gutenberg invented a printing press with metal movable type. With this single invention, the age of mass production had arrived; and following in its wake, a continuous stream of texts, records, and written materials of all kinds spread steadily into the nooks and crannies of every continent.

The second stage of the evolution of human communication involves the transition from the invention of paper and the mechanical printing press to the growth of electronic forms. As in the case of writing, advancements in electronics followed a progression that built on early innovations and advanced to higher levels of complexity and sophistication. Communication innovations paralleled the emergence of Western industrialization that began in the late eighteenth century, expanded during the nineteenth and twentieth, and shows no sign of slowing down in the twenty-first.

Communicating does not involve only texts containing words but images as well. In 1814 CE, Joseph N. Niepce developed the first photographic images. Then, in 1831 one of the most important dates in the advancement of electricity driven devices, Joseph Henry produced the first electric telegraph. Shortly thereafter, Samuel Morse made the first long distance telegraph line and used his newly created Morse code as the medium for sending messages from one location to another. From that point forward, a parade of inventors followed and overlapped each other in generating an explosion of novel forms of electronic communication in both word and picture that eventually enveloped the entire planet. New types of communication devices were invented during the decades of the 1870s and 1880s. In 1876, Thomas Edison patented the first office copying machine and Alexander Graham Bell the electric telephone. A year later Edison patented the phonograph and Eadweard Muybridge invented high speed photography—the forerunner to motion pictures. In 1888 George Eastman patented the roll film camera. Then, like innovators who centuries earlier improved the quality of paper and the efficiency of the printing press, inventors introduced improvements in virtually every area of communication.

Electronic communication was not long confined to just wires when Marconi sent radio signals across the Atlantic Ocean in 1902. Then, only two decades later in 1923 Vladimir K. Zworykin invented the cathode-ray tube (CRT) and the first television camera. By the end of the first quarter of the twentieth century the radio/TV age had been launched. This in turn stimulated the rapid acceleration of other technological innovations, such as the transistor (1948), integrated circuits (1958), and the microprocessor (1971), all of which fueled the late twentieth century's most significant communication invention—the computer.

In 1990, English computer innovator Tim Berners-Lee devised the World Wide Web. When the U.S. government released the Internet into the public domain in 1994, the age of global electronic communication took another huge step into the future. As a consequence of coupling the Internet with current innovations in cellular telephone technology, time and space have ceased to be barriers that once limited both the range and content of transmitted messages.

Since *Home-sapiens* first appeared on Earth, the quest for constant self-improvement transformed slow moving word of mouth forms of communicating into a global electronic network that moves massive amounts of information at the speed of light. None of this would have happened without one essential prerequisite: humanity's capacity and desire to create, store, and use knowledge to overcome limitations of nature and advance along the evolutionary pathway.

What drove all this inventiveness and development? Certainly, some of the most important motivations include the desire to understand how things work. There were economic motivators as well. Another example of a motivator for knowledge is the desire for people to encounter new cultures, to conduct business in new markets, and to experience different places around the globe—in a word—to travel.

Transportation

The same can be said for the area of transportation. Like the area of communication, transportation inventions gave humanity the ability to overcome the space and time barriers that impeded movement from one location to another.[5] From the time of the Stone Age until the emergence of the scientific and industrial revolution 250 years ago, the majority of travel enhancements applied to improving travel over land and water. When the Sumerians started using the wheel around 6500 BCE, it was only a matter of time before humans domesticated animals to pull wheeled vehicles, as the Mesopotamians did around 3500 BCE.

In water transportation, the Egyptians traveled in reed boats by 6000 BCE, and the Mesopotamians added sails in 4500 BCE. By 3500 BCE oar powered ships were sailing the Eastern Mediterranean seas. Improved transportation, such as the chariot, increased effectiveness in battle. Domestication of animals made supplying the armies more effective and allowed, for example, the army of Alexander the Great (356–323 BCE) to be

---

5. Herbst, *A History of Transportation,* describes the major changes that humans made in how to travel from the ancient invention of the wheel to modern space flight.

much more wide ranging. By 44 BCE, the Chinese created wheelbarrows to haul loads too heavy for people to carry.

From these bygone days of antiquity to the rise of the scientific era, most of the changes in land-and-sea travel built on the foundation of innovations that emerged thousands of years ago. With the invention of the steam engine, Robert Fulton launched his first voyage of the North River Steamboat in 1807. When George Stephenson produced the first major railway locomotive in 1814, the era of the steam driven railroad had arrived and along with it the application of this new technology to bigger and faster moving machines of all types.

The process of producing new and better ways of moving people, military weapons, and commerce from one location to another did not stop with the steam engine. In the same year that Fulton's North River Steamboat made its first cruise, in 1807 Isaac de Rivas invented the first internal combustion engine. Although his hydrogen gas powered vehicle proved to be an unsuccessful design, it open the door to the development of the gasoline engine automobile that Jean Lenoir is credited with creating in 1862. With the advent of the motor car, the next chapter in the transportation story is well known. The days of highly successful steam driven transportation were numbered.

In 1885 Karl Benz built the world's first practical car with an internal combustion engine. Then in 1908 when Henry Ford used assembly line technology to produce the world's first affordable mass produced automobile, the Model T or Tin Lizzie, he revolutionized the world of transportation. With this single manufacturing innovation, the world of travel became democratized like never before in history as ordinary citizens for the first time were able to buy a means of transportation that gave them the flexibility to go where they wanted whenever they wanted. The worldwide transformation that the internal combustion engine created during the twentieth century can be likened to the impact that the written word had millennia earlier in the realm of knowledge. Its global impact on human mobility over land and on top of, as well as below, sea surfaces cannot be overstated.

The next decisive innovation that thrust against nature's time and space barriers was the airplane. It was in 1903 that the American Wright brothers flew the first modern airplane propelled by the internal combustion engine. This helped to launch later developments in jet and rocket technologies, which have enabled humans to soar skyward to heights and at speeds that earlier generation could not even imagine.

By the end of the twentieth century and into the twenty-first, worldwide commercial air travel in near Earth space was routine. The first Moon landing in 1969 was a necessary precursor for launching even deeper space probes that stretched the boundaries of space exploration in the direction of distant planets and other objects that encircle the Sun and beyond. At present, we might speculate that we humans will venture outside the solar system using new inventions and approaches that will be needed to make interstellar travel more than just science fiction fantasy.

However, if the history of transportation innovations, especially during the twentieth century, offers any clues to the future, then humanity will follow in the footsteps of ancient ancestors who first invented the wheel. The desire to expand our reservoir of knowledge will continue to drive us and our robotic probes toward still unexplored worlds. In parallel to the impulse to communicate that led to writing and the global Internet, there will be no retreat from the quest to invent novel ways to overcome the natural barriers that block the human urge to travel to the remote fringes of deep space whenever, wherever, and however—without limitations—in a word, to the planets and the stars.

Planets and Stars

From the beginning, we humans have looked skyward with wonder and awe at the magnificence of the starry night. Questions concerning the origin of the plants and stars spawned many diverse and colorful images of the cosmos. When they are placed side by side, they separate into two distinct branches. The first consists of a collection of creation stories that appear in the diverse cultures and religions prior to the rise of modern science. The second involves a view of the cosmos that emerged during the modern scientific era.

Among the pre-scientific views, three are well known. First, in Western culture, the dominant outlook on the origin of the universe is found in the first two chapters of the book of *Genesis* of the Jewish Scripture that Christians call the *Old Testament*. God created the universe in seven days through divine command along with all creatures, including humanity— the crown of creation—and inanimate objects that exist within it. No other universe exists. Not only did God bring the cosmos into existence but God will also bring it to an end at a time that only God knows. In other words, the God who created and remains separate from, yet interactive with, the cosmos is both the *alpha* and *omega* of the Judaic-Christian creation story. History is linear, it does not recycle and there is no reincarnation.

Second, in South Asia, the Hindu story of how Lord Brahma created the universe dates back many centuries. The world that currently exists is not the first universe nor will it be the last. There have been and will continue to be as many universes as there are drops of water in the Ganges river. Time is cyclical, and universes come and go. When a new cosmic cycle starts, the material universe emerges out of and becomes the physical body of Lord Brahma, whose spiritual presence remains veiled within it.

A third popular pre-scientific view of the origin of the universe stems from ancient Chinese lore. It involves the interplay between the two great energies of the cosmos—the *yin* and *yang*. The *yin* is soft, female, and receptive; and the *yang* is fiery, male, and energetic. Together they produced the universe and all the finite beings that inhabit it. Harmony prevails throughout the whole when balance is maintained between these two forces.

In addition to these three well-known ancient accounts of creation, there are many others that appear throughout the world's diverse cultures.[6] It is not the intent of this chapter to include all of them. At the same time, with few exceptions, one overarching aspect of virtually all pre-scientific views of the cosmos is that in one form or another they looked on the Earth as the center of the universe with the Sun, Moon, planets, and stars circling around it. This is reasonable since; after all, that is what we observe. In the West, this view was advocated by the Greek philosopher Aristotle whose ideas reinforced the *Genesis* creation story. Building on these two ancient authorities, in the year 200 CE, Ptolemy proposed an Earth centered view of the universe with objects in space moving in perfect circles around it. It was assumed that everything beyond the Earth was perfect. Circles were perfect. Therefore, all objects beyond the Earth moved in circles.

This view prevailed for centuries until around 1500 CE when those who sought after knowledge and understanding began to question this longstanding, taken for granted conception of the cosmos. They were willing to question what was accepted for 1300 years. Then, in 1543, the amateur Polish astronomer and mathematician Nicholas Copernicus

---

6. For an excellent summary of the world's many creation myths, see Leeming and Leeming, *Encyclopedia of Creation Myth*. Leeming and Leeming, A *Dictionary of Creation Myths*.

(1473–1543) made one of the most important scientific conjectures ever. Based on careful observations of the stars and planets (taken without a telescope), he realized that if the Sun was the center of the solar system and not the Earth, his observations could be better and more simply explained. This explanation marks the beginning of modern astronomy and, for some, of modern science itself. It opened the door to a new wave of astronomers whose findings in successive generations confirmed Copernicus's original premise.[7] Copernicus did not depend on ancient authorities whereas Ptolemy did. Copernicus relied on this observations and his explanation, which was simpler than the convoluted Ptolemaic model. Compared to Ptolemy's theory, Copernicus's theory was simpler, which is usually preferred for explanations of our natural world, all else being equal.

During the year 1609, two astronomers in particular made discoveries that not only supported Copernicus's Sun-centered view of the solar system but took it to the next level. The first is Galileo Galilei (1564–1642) who demonstrated by observing the moons of Jupiter with his new and more powerful telescope that not all objects in space encircle the Earth. He also predicted that Venus would exhibit a full set of phases like the Earth's encircling Moon, which supported Copernicus's idea that the planets revolve around the Sun. The second is Johannes Kepler who further undermined the Earth centered view by observing that the planets move in ellipses around the Sun and not in perfect circles. The long held fallacy was that all bodies above the Earth were perfect since they were in the heavens and moved in perfect circles. Like Galileo who observed imperfect Sun spots, Kepler conjectured that planets moved in imperfect ellipses, which better predicted their motion. Since improved predictability is a key feature of successful scientific theories, Kepler's successful law of planetary motion became the basis for later discoveries.

In 1687, Isaac Newton developed his three laws of motion and the Law of Universal Gravitation that lies at the heart of classical physics and serves as the basis for space travel. By the early 1900s, astronomers began measuring distance from the Earth to the Sun as well as to other stars. Then in 1905 and 1915 respectively Albert Einstein published his special and general theories of relativity out of which emerged a lively debate over whether the universe is finite or infinite. Einstein argued for a Cosmological Constant that implied space is finite. However, Edwin Hubble (1889–1953), one of

7. For a detailed account of the discoveries of Copernicus and subsequent astronomers, see Hoskin, *The Cambridge Concise History of Astronomy.*

the twentieth century's most important cosmologists, in 1929 discerned through his observations and analysis of intergalactic relationships that the universe is not static. On the question of whether the universe is enlarging, contracting, or remaining static, Hubble concluded that it is expanding.

Prior to 1950, astronomers disagreed over whether the universe along with its size and properties had always remained in a steady state condition or whether it had a specific starting point in time. The well-known British astronomer Fred Hoyle was a strong advocate of the steady state view. However, the accumulated findings from Earth based telescopes and from satellites pointed to another explanation. In 1950 Hoyle dismissively labeled his opponents' view as the Big Bang. By 1970, just twenty years later, almost all observational cosmologists had come to reject Hoyle's steady state theory. At the same time, in a twist of poetic humor, they retained his colorful image of a universe that began in a hot and fiery explosion—the Big Bang—that erupted 13.7 billion years ago and has been expanding ever since.

Thus, from the earliest pre-scientific speculations about the origins of the cosmos to the modern scientific view of the Big Bang, we humans have relentlessly pushed the boundaries of our knowledge closer to the outer edges of the universe—literally. This has been expensive. Here are three examples of actions we have taken in the continuous quest to expand our knowledge:

- Large Hadron Collider. This is the largest, highest energy, and most expensive subatomic particle accelerator in the world (started in 2008). Its total cost is about $10 billion. It was built to advance human understanding of the deepest laws of matter and energy and how they relate to the Big Bang. These laws relate to the most basic structure of matter and related fields, energy, and how they are related. A summary of this work would quickly take us way beyond the scope of this book.[8]

- Hubble Space Telescope. This is a space telescope that was placed into orbit around the Earth in 1990 and remains functional at the time of this writing. Many Hubble observations have led to breakthroughs, such as accurately determining the rate of expansion of the universe and associated dark energy.

8. "Large Hadron Collider."

- Curiosity rover on the surface of Mars. A car sized rover was placed on the surface of Mars to determine (among other goals) if that environment might have been favorable for the development of microbial life.

Each of these three examples has little, or no, value for our immediate survival. The expense, the technology development, and the engineering were designed to reach out and answer basic questions about the composition of our universe, the possible origins of life, and the most fundamental structure of matter.

Why do we do this? We want to know how the universe got started, where it is going, where we fit into it, and what our ultimate outcome will be. Despite the costs and lack of practical applications, scientific understanding was and continues to be the goal. When we couple our current and improving communication and transportation knowledge and technology with our modern comprehension of the cosmos, will we one day travel outside our solar system to the distance stars we know are there?

Health and Longevity

The last area to be included in this chapter involves knowledge that relates to maintaining the health and longevity of the human body. It would not be an exaggeration for us to say that nature's limitations in this area are as formidable as in any other. We are reminded continually of the fragility of being human in and through the events of everyday life. Despite the body's self-sustaining vitality and remarkable vigor in recuperating from infirmities of all kinds, everyone everywhere eventually realizes that human life is finite. Our bodies do not last forever, and they all wear out and die. Furthermore, when compared to the estimated age of the universe since the Big Bang—13.7 billion years ago—the span of any person's brief time on Earth is incomprehensibly small.

In addition to the universal awareness of our finite human nature, the old adage that "life is not fair" takes the restrictions of nature to another level. Sooner or later everyone learns that in the gap between birth and death, life does not distribute its burdens and benefits equally. Life is filled with whims and inconsistencies. Some people are born with abundant blessings, and others start with very few. This is especially true in matters of health and longevity. At every stage of the life cycle, every individual is subject to illnesses and fatalities that frequently appear to be random and can strike at any moment. Death comes early, and often tragically, to some and not to others. Being born does not carry with it a guarantee of a long and prosperous life. In due course—without exception—everyone recognizes that the potential ravages of disease and dying can affect anyone at any age at any time and in any number of circumstances.

It is this realization that has motivated humans to search for ways to cure and conquer some of the harshest conditions that nature imposes on the human body.[9] The timeline of advances in treating disease in Western culture dates back to the Greek physician Hippocrates, born in 460 BCE. His embryonic and experimental treatment practices laid the foundation for gradual improvements that emerged slowly during the following millennia. In the first and second centuries BCE, as a result of his care of several Roman Emperors and scores of injured gladiators, the physician Galen wrote over 500 treatises that offered new knowledge in the areas of anatomy, physiology, and pharmacology. While their knowledge of how the body operates was often inaccurate, their lasting contribution stems from their method, which they held should be based on objective observations and inductive reasoning.

This ancient legacy of learning that started with Hippocrates continues to the present day and underpins all modern scientific approaches to examining how the body's many systems function separately as well as interact together. Anyone who spends even a modest amount of time studying the history of medical advances will discover quickly that nature did not yield its secrets easily. As in the case of communication, transportation, planets and stars, and later innovators in medical research and practice always stood on the shoulders of those who came before them. In some cases new knowledge led to discarding old ideas. In others, it emerged by updating them. Either way, persistence paid off as precise knowledge of the body's operations started to surface.

This would not have been possible without the development of novel technologies that enabled researchers to discover the internal structures that lie hidden beneath the body's outward appearance. When the Dutch lens grinder Zacharias Jannssen invented the microscope in 1590, he had scant awareness of how this single innovation would revolutionize medical practice. Eighty years later in 1670, after experimenting with 500 different microscope models, Anton van Leeuwenhoek discovered blood cells and other types of cells while inspecting plants and animal tissues. These and

9. Porter, *Blood and Guts.*

subsequent inventions led to expanding the sphere of ever more precise knowledge of the body's microscopic milieu.

In turn, this laid a foundation for the founder of modern immunology Edward Jenner, who initiated in 1796 one of the most significant interventions in medical history. He scratched the arm of his gardener's eight year old son with pus that contained the cowpox virus and then exposed him to smallpox six weeks later. His expectation that the body's reaction to the cowpox virus would create resistance to smallpox proved correct. Starting from this single experiment, the practice of vaccinating against smallpox (from the Latin word *vacca*, which means cow) grew rapidly. By 1980, Jenner's unprecedented accomplishment had swept across every continent when the World Health Organization announced the global eradication of this dreaded disease.

The next major step forward in expanding the practice of immunization occurred with the development of the germ theory of disease, which is arguably one of the humanity's most significant scientific discoveries ever to help in surmounting some of nature's worst biological ravages. Credit for establishing the germ theory during the 1870s goes to Louis Pasteur and Robert Koch. Their convincing demonstration that specific organisms cause specific diseases created a paradigm shift in medical thinking that spawned the development of dozens of new vaccines.

Since 1879, highly effective vaccinations have developed for over twenty-one major diseases such as cholera, rabies, and hepatitis A. In a span of slightly more than two hundred years from 1796 when Jenner discovered a vaccine for smallpox until the present day, practitioners began applying modern inoculation procedures to prevent the outbreak of destructive diseases that theretofore had victimized millions of people for centuries.

With the repeated scientific confirmation of the germ theory as demonstrated by the growing practice of immunizing against deadly diseases, it was only a matter of time before innovations emerged in other areas as well. Paralleling the advancement of vaccines, researchers began looking for ways to cure a wide variety of germ related afflictions. These initiatives piloted the development of antibiotics—also called wonder drugs.

Building on earlier laboratory experiments, Alexander Fleming demonstrated in 1928 that a type of penicillin mold could destroy staph bacteria without failure or toxic side effects. Fourteen years later in 1942, when Howard Florey and Ernst Chain started selling the penicillin drug to the public, the mission to create a broad range of antibiotic drugs capable of killing many kinds of bacteria although not viruses accelerated. One year after penicillin became available for widespread use Selman A. Waksman in 1943 discovered streptomycin for treating tuberculosis and an array of other life threatening diseases.

The story line on the search for cures that would enhance human health and longevity continued without interruption. In 1955 Lloyd Conover patented tetracycline, which became the most widely used broad spectrum antibiotic in the U.S. When SmithKline Beecham patented amoxicillin in 1981, the era of semi-synthetic antibiotics began. Despite the growing resistance of a number of bacteria to certain antibiotics during the past two decades, the public demand for newer and more powerful wonder drugs continues unabated as does the competition among drug companies to develop them. The old aphorism "Nothing succeeds like success" finds ample support in the human drive not only to push back disease boundaries but, if possible, to eliminate them altogether.

It is impossible to overstate the positive effect that the knowledge and application of vaccines and antibiotics has had in helping humanity overcome some of nature's most devastating diseases. At the same time, this is only part of the story. Parallel progress in several other areas of medical intervention has contributed in equal measure. While it is not the purpose of this chapter to detail all of these achievements, some are especially noteworthy because their impact has been truly transforming in enhancing health and extending human longevity.

For example, since the invention of the stethoscope by Rene Laennec in 1816, medical advances in listening to and visualizing in great detail the body's internal structures have led to the development modern x-ray and MRI (magnetic resonance imaging) techniques. In 1901, when Karl Landsteiner classified blood into four compatibility and incompatibility groups of A, B, AB, and O, he set the stage for performing his first successful blood transfusion in 1907. This was followed by the establishment of the first blood bank in Chicago in 1937 by Bernard Fantus.

These two innovations paved the way for Paul Dudly White who began using the electrocardiograph in 1913 with the goal of finding new paths to improve heart health. In 1935, John H. Gibbon, Jr. employed an innovative heart-lung machine while he conducted heart surgery. Less than twenty years later, in 1952 Paul Zoll stabilized an irregular heartbeat by producing the first pacemaker. By 1967, the cumulative advancements in cardiologic knowledge and technologies enabled South African Christian Barnard to perform the first human heart transplant. The first artificial heart operation took place in 1982 when William DeVries implanted the Jarvik-7 in Barney Clark who lived for an additional 112 days. Since then, open heart surgery to repair obstructed arteries and veins has become routine. When we recognize that use of the electrocardiograph dates from 1913, progress in enhancing heart health and longevity during the past one hundred years has been nothing short of astonishing.

In addition, other discoveries and innovations could be added to the record. While many of these led to refinements in existing procedures, others opened entirely new areas. For example, for infertile couples, human embryos called test tube babies can now be conceived artificially in a laboratory through the technology of *in-vitro* fertilization and then transplanted successfully into a woman's body. In 1996, Dolly became the first cloned mammal from adult sheep cells. Since then, the practice of cloning other species of animals has become routine. Breakthroughs in stem cell research in the 1990s open up the possibility of treating and eradicating disabling diseases that appear later in life as a result of genetic impairments that are inherited at birth.

While these and other innovations have led to improving health and longevity during the past half-century, two of them in particular tower above the rest—describing the shape of the DNA molecule (deoxyribonucleic acid) and mapping the human genome. By the early 1950s, scientist knew that chromosomes carried the body's genes or units of inheritance called the genetic code. Using the x-ray investigations of Rosalind Franklin and the research of Maurice Wilkins, James Watson and Francis Crick were the first scientists to accurately model the twisted strand of the tightly coiled DNA structure known as the double helix.

The DNA discovery laid the foundation for the Human Genome Project, an international research initiative whose goal was to determine how approximately 23,000 genes that direct human growth and development are sequenced on the body's chromosomes. The Project began in October 1990 and was expected to be completed in fifteen years. Ten years later in 2000 a working draft of the genetic map was created. In April 2003 knowledge of the human genome was declared complete. Through the concerted efforts of many scientists, led by J. Craig Ventner and Francis Collins, the Human Genome Project achieved its objective two years ahead of schedule.

There are many commentators who look upon discovering the structure of DNA and sequencing the human genome as the most significant scientific breakthroughs since the time of Hippocrates. When the working draft of the human genome was published in 2000, then President Bill Clinton announced at his June 26 news conference (flanked by both Ventner and Collins) that "Without a doubt, this is the most important, most wondrous map ever produced by humankind."[10]

While such praise sounds excessive, it is not—for one reason. For the first time in history, humanity now possesses a form of knowledge that could lead to removing biological barriers that were once thought to be permanent and beyond which no further improvements in human health and longevity could go. In particular, the area most immediately affected involves genetic modification with long term implications for life extension.

For example, as a result of enhancements in the fields of modern medicine, public sanitation, and genetics during the last one hundred years, average life expectancy in the US has increased from age fifty seven to seventy eight. Other developed nations have also improved the average life expectancy of their citizens. As knowledge of the human genome is applied to the progressive elimination of genetic diseases,[11] it will contribute to the continuation of this trend throughout the developing areas of the world as well.

However, average life expectancy is not the same as the maximum life span with its current outer biological boundary of 120 years. The question of whether we humans will use our knowledge of the human genome to push the current barrier past 120 years remains to be answered. While much debate rages around this issue, some biological gerontologists like Aubrey de Gray[12] foresee that humans will reach 150 in the not too distant future and eventually will live for 1000 years—maybe more, maybe forever. Others see this kind of conjecture as mere fantasy bordering on science fiction with scant support given the present state of genetic science.

10. Emmett, "The Human Genome."

11. W. French Anderson was one of the early pioneers in the field of gene therapy. In 1990, he replaced a defective gene (known as ADA that causes a severe immunodeficiency in the body) in Ashanthi deSilva, a four-year old girl. When she turned 21 in 2007, she was still stable from his genetic modification intervention. Since 1999, the science of altering defective genes inherited at birth has proceeded with great caution following to the tragic case of Jesse Gelsinger who died unexpectedly from an experimental treatment that involved faulty protocols. See Munson, *Intervention and Reflection*, 123–27.

12. De Gray and Rae, *Ending Aging: The Rejuvenation Breakthroughs That Could Reverse Human Aging in Our Lifetime*.

While the life extension movement is still in its infancy and concrete evidence for extending human life beyond 120 years has not yet been forthcoming, the scientific mapping of the human genome along with research on stem cells serve only to fuel such speculations. Why this should be so is clear. As in the areas of communication, transportation, planets and stars, knowledge of how the human body functions has grown steadily since the rise of science during the modern era. We humans have searched for centuries to discover ways to improve health and live longer; and now we stand on the edge of a new era where knowledge of the human genome has emerged as the holy grail for combating and possibly even conquering nature's ultimate biological barrier—death itself.

Conclusion

Several conclusions and well as questions can be drawn from the above discussion. First of all, it is clear that through incredible advancements in knowledge, we humans have pushed back, and even overcome, many of the barriers that nature has placed in front of us during the 200,000 years we have been here. In the area of communication, we have progressed from limited face to face talking in early tribal settings to global, speed of light electronic conversations without geographical restrictions. Changes in modes of transportation followed a similar pathway. Early humans spread out across the land on foot. Now we soar into outer space with Moon landings, planetary fly bys, and excursions with rocket propelled spacecraft.

Ancient musings about the starry skies and the pre-scientific creation stories they inspired gave way to the modern Big Bang theory of the origin and expansion of a vast cosmos filled with billions of stars that inhabit billions of galaxies. Our ancestral fascination with this heavenly canopy has continued without interruption down to the present day.

The struggle to contain and eliminate deadly diseases emerged early in human history. Building on this bygone legacy, modern medicine, which has achieved such towering heights of understanding the human body, would leave the ancients in awe. Innovations in stem cell research coupled with decoding the double helix of DNA and mapping the human genome have pushed health horizons to unprecedented levels. For the first time in human evolution, we stand on the threshold of not only eliminating once thought to be incurable genetic diseases but of extending life itself beyond the present boundary of 120 years.

Given the relentless pattern of persistence in pursuing new levels of knowledge that have led to progressive improvements in the human condition, we are left to wonder what is next. The answer, of course, is that we do not know, especially in the long run. However, what we do know is that short of a dramatic and unexpected reversal of global trends that have emerged during the past two hundred years, particularly since the rise of modern science, we will continue to discover domains of knowledge that will take us into new and still to be explored frontiers.

In the following chapters, we will examine in greater detail several of the topics highlighted in this chapter as well as many others that we have not yet discussed. In turn, this will lead to the final chapter in which we will take up the case of whether the cumulative evidence based on advances in scientific knowledge increases or decreases confidence that God exists.

3

The Universe Is Structured
for Conscious, Self-Aware Life

Introduction

All of the world's major religions have creation stories that explain how our universe came into existence. Most all of us desire to have some understanding of such beginnings and there are religious responses to our longings.

Taking into account current scientific explanations for how our universe started, at the most fundamental level there are but two options that address our questions about origins: 1) Our universe was initiated by random and undirected causes and origins and 2) our universe came into being as a result of purpose and design. Using current scientific conclusions, in this chapter we will identify and detail these issues with the goal of acquiring a higher confidence level for one of these two options. We will draw our conclusions in chapter 8.

As a start let, us think of the universe at the beginning, 13.7 billion years ago, at the Big Bang, and also think of the universe today when we have conscious, self-aware humans. This is summarized in Figure 2. How did our universe get started? What was its cause and origin? Of course we cannot go back in time and make definitive observations. We must rely on the current best evidence that might indirectly contribute confidence levels to these really fundamental questions. Notice in Figure 2 the role of two types of evolution, one linked to life and the other linked to the physical. These two types overlap and in chapter 8 we will see some unexpected parallels.

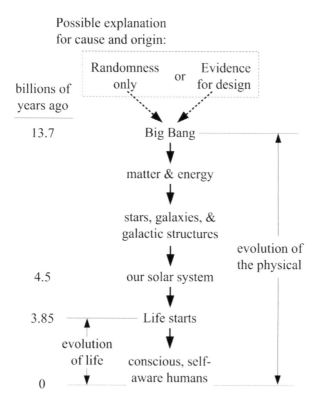

Figure 2. Simple timeline for the period between the Big Bang and today.

There is much relevant and currently accepted information related to these two options (randomness and design) from the fields of cosmology, particle physics, astrophysics, earth sciences, biology, anthropology, and chemistry to name but a few. After all, we span 13.7 billion years here and that is a long time. How can we possibly make sense of all of it? How can this information help us address these questions about origins? The answer is that we need some kind of tentative explanation, a theory that we can use to compare our observations, our simulations, and our other data. This is similar to Darwin's theory of evolution where his explanation for the diversity, linkages, and development of life involved life forms that successfully found niches in local ecosystems. This allowed for those life forms to compete and survive against other local life forms that were less fit. Comparisons could then be made and observations and predictions could be tested. We seek a similar kind of theory.

What kind of theory or explanation might we use to guide us in developing more confidence in one of these two options, randomness or design? We believe that one of the strongest theoretical candidates is the Anthropic Principle.

The Anthropic Principle

The Anthropic Principle[1] was first introduced in 1973 by the astrophysicist and cosmologist Brandon Carter from Cambridge University, at a conference held in Poland to celebrate the five hundredth birthday of the father of modern astronomy, Nicholas Copernicus. The Anthropic Principle involves identifying evidence that seems to correspond to the observation that our universe was destined to evolve into the type of life we experience all around us (carbon based life, dependent on water)—that the universe has to be a very special place. The laws of nature in such a place must take particular forms. All the forces of nature must have particular strengths. Not only that, but there are very narrow margins (plus or minus ranges or windows) that are allowable for these force strengths.[2]

While the universe did not have to evolve as it did given other possible mathematical and physical combinations, nonetheless, it has. If our universe did not have these strengths and their associated allowable and narrow windows, then our type of life would not have come into being; and we would not have evolved as we have within the lifetime of our universe.[3] Stated differently, the Anthropic Principle refers to the seemingly arbitrary constants in physics that have a mysterious and common characteristic: These are precisely the values needed to produce and support life.

Among the astronomers who study the origins of the cosmos, there is widespread consensus that the universe started 13. 7 billion years ago with the Big Bang explosion that released unimaginably huge amounts of matter and energy. At approximately that time and location there were

1. The version of the Anthropic Principle we use here states that the laws and constants of nature must be as they are in order for our type of carbon based, water dependent life to exist.

2. Stannard, *Science & Belief*, chapter 5, "The Anthropic Principle."

3. Polkinghorne, *Theology in the Context of Science*, 100–107.

incomprehensible temperatures and energy densities. Here is a current conclusion from the US Department of Energy and National Science Foundation: "These processes had to be finely tuned to yield a universe capable of forming the galaxies, stars and planets we observe today. Did some undiscovered fundamental laws determine the conditions that allowed us to exist?"[4]

This question can be answered only in part but not in full as the following summary of breakthrough discoveries indicates. Over the past thirty years, scientists have developed a deeper understanding of the natural laws that have a controlling influence on time, energy, and space. Based on numerous experiments that have been conducted in laboratories around the world, an array of research teams have been testing the basic assumptions of what is called the Standard Model. This Model is nearly universally accepted because predictions based on its premises have held true time and again. "The series of experimental and theoretical breakthroughs that combined to produce the Standard Model can truly be celebrated as one of the great scientific triumphs of the 20th century."[5]

However, despite the predictive potential of this Model, a fundamental mystery surrounds it. At this time, we have no explanation for it despite its straightforward and comprehensible nature. As it currently exists, the Standard Model is pretty simple. It is a nice and tidy matrix of particles—only six quarks, six leptons, and four Gauge bosons—that describe our whole known universe. The mystery is this: Given the incomprehensibly vast and turbulent explosion of the Big Bang, why is the Standard Model so simple and why does it have such broad application?

In association with the Standard Model, we have additional understanding of how the universe functions. For example, there are only four forces that operate in our cosmos. They are 1) the weak and 2) the strong nuclear forces. Because these are active only on the scale of the atomic nucleus, they are not part of our common everyday experiences. 3) The third force is Electromagnetism that is responsible for x-rays, light, radio communications, and many other aspects of our lives such as the hardness of materials. 4) Gravity is the fourth force, which will be discussed in detail in the next section of this chapter. Like the Standard Model described above, we are left wondering how this seemingly infinite, complicated, and

4. "Quantum Universe."
5. "Quantum Universe."

amazing universe can be explained with such apparent simplicity that includes only four forces.

Other discoveries have led to additional types of mysteries. In 1928, theoretical physicist Paul Dirac joined the mathematics for Einstein's theory of relativity and new discoveries in quantum mechanics. This led him to predict the existence of antimatter, a completely different form of matter. A particle of matter, such as an electron, has a corresponding antimatter particle, such as a positron. The masses and energies of the two particles are equal but the charges are reversed. When they come together they can annihilate each other, releasing all of their combined mass as energy, say, as gamma rays. Four years later in a laboratory experiment, antimatter was duly discovered and predicted by Dirac's mathematics and interpretation. The mystery here is this: Why should the mathematics of Dirac, which after all is merely the product of brain electro-chemistry, combine to describe this surprising physical reality, which was not anticipated or known until that time? While it is not self-evident that there should be any correspondence between the simplicity of mathematics and the laws of the universe, in fact there is as the presence of antimatter throughout the entire cosmos reveals.

At this point, a word of cautious is in order. Lest we think that we have most of the universe figured out, we do not. To date all of our observations and widely accepted conclusions related to the Standard Model, the four forces, and antimatter are at best incomplete views of the universe. Despite what we do know, there is much that we do not know. For example, about 95 percent of the universe about which we know nothing is not made of ordinary matter. It still remains a mystery that has been defined only by the terms "dark matter" and "dark energy." In other words, despite what we do know, today we really do not know much about the basic substance that comprises most of the cosmos. Perhaps we will discover additional order in this 95 percent. See Figure 3.

Nonetheless, despite the existence of many unanswered questions that expose our limited knowledge of the universe, there is a great deal that we do know. In the rest of this chapter, we summarize this knowledge. Building on the above definition of the Anthropic Principle, we will start with a discussion of the importance of gravity, the last of the four forces explained above, and then consider the issue of randomness versus design as it applies to the origins and nature of the cosmos.

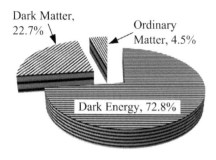

Figure 3. The proportions of dark energy, dark matter, and ordinary matter in the universe.

Gravity

One of the best examples to illustrate how the Anthropic Principle appears to operate in the universe is gravity. Gravity is a force we feel in our everyday lives. It is also the force that holds GPS satellites in orbit around the Earth, which help us with directions to the home of an out of town friend. It constrains the planets to orbit about the Sun. It holds our Sun within our galaxy. Gravity is both a localized force and a force that extends across the universe. It is a force that operates on the scale of our everyday world as, for example, when a pencil drops to the floor as well as on a completely larger scale of galaxies and clusters of galaxies. We know that gravity spreads out in a certain fashion, and it has a specific strength. Distance from the object, its mass, and something else set the gravitational force. There is a single number that describes gravity's strength on the scale of our everyday lives all the way up to the scale of galactic clusters and beyond. This number (the something else) is called the gravitational constant.[6] It is astonishing: this one number applies to our everyday lives as well as to stellar formation, to galaxies, and to galactic clusters. All of these properties of gravity are observable, predictable, and can be described with mathematics.[7]

---

6. The gravitational constant, G, is 6.7 x 10-11 N (m/kg)2. General Relativity supersedes Newton's Law of Universal Gravitation but for our purposes here we can use Newton's law without loss of accuracy.

7. Gravity is one of the least understood forces and might rely on the Higgs boson. Here we are referring to Newton's Universal Law of Gravitation.

If this gravitational constant was slightly smaller and gravity was a little weaker than what exists, life on Earth would have developed differently, if at all. For example, birds and other life that fly would likely be more dominant. We humans would probably be taller, assuming we developed at all. The atmosphere would less tightly be bound to the Earth and it almost certainly would be thinner than what we breathe today, which would favor complex life with larger lungs.

The Earth would have formed differently, compared to what did happen, starting 4.6 billion years ago. The Earth would likely not have the current stabilizing effects of the Moon, which would mean more radical and quick climate changes that make the emergence of intelligent life much less probable. The Earth, now with a different size and volume, would likely form at a different distance from the Sun, probably further away. In this case, winters would be longer and summers colder. Assuming that humans could even develop in these conditions, we would have rounder bodies since this colder climate would favor bodies that are more efficient at retaining heat.

The temperature of the Sun, influenced strongly by gravity (and thermo-nuclear fusion), would burn at a reduced temperature. Atomic elements that evolve in this nuclear furnace would form at different rates, if at all. The planets were formed from the leftovers of previous stars. When they exploded (one or more as nearby super nova) there would be a different distribution of atomic elements compared to what currently exists. It is likely that a small change in this distribution (due to the small change in the gravitational constant) would have a major effect on how life formed and its evolutionary pathway, summarized in chapter 4. It might be, for example, that with such a small change life would have developed with no more complexity than single cell bacteria; hardly intelligent, self-aware beings. In short, the allowable range of values for the gravitational constant is small.

In addition, if the Sun would burn at a reduced temperature, there would have been an important secondary effect on the evolution of life. About 530 million years ago there was a rapid appearance of many major animal groups as evidenced from the fossil record, accompanied by major diversification of life. This is called the Cambrian explosion. Before about 580 million years ago, most of life was simple, comprised of individual cells and less frequently colonies of cells. During the next 70 or so million years, the evolution of life accelerated by about ten times. The variety of life began to be like what exists today.

Why was there this explosion? One explanation is that in animals the eye developed at this time. Once a predator like a trilobite has visual sensitivity, it has a huge evolutionary advantage over its prey without eyes. The quarry of the trilobite must have felt a powerful selection mechanism to also develop eyes to help escape hungry trilobites. This was not the only evolutionary pressure but likely an important one. As a result the predators must have developed more sophisticated hunting capabilities. To keep pace, the quarry must also have developed more sophisticated defense mechanisms to survive. As observed in the fossil record, an evolutionary arms race "soon" arose, which favored the development of more complex life forms.

If the Sun burned at a reduced temperature the sunlight would have different colors and the energy output of the Sun would be reduced, which would not have matched up so well with the development of vision about 530 million years ago (assuming the development of the eye played a key role at this time). The eye converts light energy at certain colors into electrochemical energy that is interpreted by the brain as images. This conversion requires a certain matching of the Sun's colors and the Sun's brightness into the electro-chemical reactions within the eye. With a reduced temperature Sun life forms may have remained much more primitive than what was produced by the Cambrian explosion. With such a Sun it probably would have taken longer, if at all, for organisms, reptiles, birds, mammals, and self-aware people to develop.

The distribution of matter in our galaxy would be different with a small decrease in the gravitational constant (for example, -1 percent). This different distribution would be quite unpredictable since it would depend on the locations of other stars, on the rates of stellar aging and resulting end of life events, and the distribution of super nova explosions. One consequence of a slightly reduced gravitational constant is quite certain: the distribution of matter that would eventually form our solar system would be much different than what actually happened. This difference would cause dissimilar planets to form, each with separate material distributions, all orbiting a Sun that is burning differently. The window of life giving temperatures and radiation might not have a planet within that zone. If a planet existed there, it might not have the right distribution of materials, atmosphere, and solvents to initiate the formation of life and to have suitable environmental conditions for the eventual development of intelligent and self-aware life.

Currently, one of the most interesting areas of astronomical research is the successful detection of planets around hundreds of stars other than our Sun, known as exoplanets. Technology has developed to the point where the secondary effects of exoplanets can be sensed.[8] Masses of these exoplanets currently fall into the range that varies from a few times that of Earth to twenty-five times or so that of Jupiter. Assuming that liquid water must be present for life to start, there must be a certain habitable zone around each star that depends on its total energy output. There is at least one star where there seems to be at least one exoplanet within its habitation zone.

If we assume that there is a life-friendly planet in this zone and that it does have the right distribution of materials, then a special kind of atmosphere is required. There needs to be an atmosphere thick enough to protect the life forms from bombardment of rocky debris within the solar system and from the solar wind, as described below. Without such protection the planet Earth would look something like the surface of Mars. Impacts from such rocky debris would tend to disrupt development of life—if not completely cause extinction. There are examples of this type of disruption in Earth's geological record, as will be summarized in the next chapter. With an atmosphere that is too thick and/or has the wrong properties heat can build up where life is developing due to greenhouse gases. This is almost certainly what happened on Venus where surface temperatures are far too hot to sustain life. Furthermore the elemental distribution of the atmosphere must be compatible with sustaining and developing life.

The Sun (and apparently most stars) produces what is called the solar wind. This wind consists of charged particles that leave the Sun's surface at nearly the speed of light. Some of these charged particles are aimed directly at the Earth. If the solar wind was able to reach the surface of the Earth (where life is developing and flourishing), it would be quite damaging to the affected life forms and it may even sterilize the whole planet. Among other outcomes, the genetic molecules would be damaged so that an array of adverse effects would likely occur such as excess genetic mutations, cancers, death, and deformities.

Some sort of protection is needed from this solar wind so that life can develop and flourish. Part of that protection is provided by the Earth's magnetic field. The trajectories of charged particles from the Sun (solar wind) are affected by magnetic fields. The Earth's magnetic field essentially reflects and/or deflects the charged particles of the Sun's solar wind so that

8. Cox and Cohen, *Wonders of the Universe*.

life on the Earth's surface remains mostly unharmed. While most, if not all, stars probably produce some kind of solar wind of charged particles, a life developing planet would require protection from this solar wind.

Probably the best way to shield the planet is through the establishment of magnetic fields. For such a magnetic field to exist, electrical currents must be present in the planet's center, implying a molten core. It is highly unpredictable whether or not such a planetary core would be produced with a sustainable and strong enough magnetic strength to protect life from the star's solar wind. However, this is precisely what has occurred in the Earth. Unlike other planets, the Earth's magnetic field, as produced by its liquid iron core, is tuned to protect life from the harmful particles that emanate from the Sun.

Furthermore, slight deviations from these current gravitational balances would have led the Earth to evolve in a decidedly different direction. If the gravitational constant was slightly larger (for example, +1 percent) life almost certainly would have developed differently, if at all. Life forms not so susceptible to gravity would likely tend to dominate such as snakes, centipedes, and sea life. Birds and other life forms that fly would likely be less important. These changes imply that most of the food chains for ecosystems would be much different, which would greatly and unpredictably affect the development of all forms of life. Possibly humans would develop. If so, we humans would certainly be much different. Our bodily structure would support different weight distribution. Our food would certainly be different, having developed from different life forms. Our heart and circulatory system would be more robust to lift and move all that blood (or equivalent) around our bodies. Muscular development would likely take priority over brain development.

The Sun would burn at a higher temperature due to the increased gravitational force. Heavier atomic elements would evolve at a faster rate. The Sun would burn up its available thermo-nuclear fuel faster. If everything else stayed the same (which, of course, it would not) there would be less time for intelligent, self-aware life to evolve. With the current gravitational constant it took 3.85 billion years for humans to develop. If the constant were slightly larger, there might be only 2 billion years or less before the Sun burns out. Probably this would not be enough time to develop conscious, self-aware life like us. The other three considerations summarized above, 1) a different distribution of matter in the galaxy, 2) special atmosphere, and 3) solar wind that apply to an increased gravitational force also apply in a similar way to a decreased gravitational constant.

Gravitational forces spread out in a manner that affects life development. Thermo-nuclear fusion in stars produces light and other forms of energy; which is balanced by gravity due, in part, to how gravity spreads out. This specific type of spreading translates into gravitational force fields extending across our bodies, eco-systems, planet, solar system, galaxy, galactic cluster, and our entire universe. Small changes in the type of spreading will have huge consequences for how life would have developed as summarized earlier in this section.

In the Beginning

Over the last forty years or so, observational astronomers have studied the early moments when the universe began. Based on data from the Hubble Space Telescope, for example, we now have a clearer picture of how this happened with many interesting conclusions.

Matter and antimatter are similar in every way except that the energy equivalent of matter is equal and opposite to the energy equivalent of the same quantity and type of antimatter as explained earlier in this chapter. Subatomic particles such as protons and neutrons consist of quarks, which are the fundamental building blocks of matter, according to our current understanding. Just as there are antimatter equivalent particles for the electron and proton, there are also anti-quarks.

Right after the Big Bang, matter and antimatter appeared in nearly equal amounts. An unanswered question today is this: Why were not the amounts exactly equal and why does this small inequality matter? As far away into the universe as can be explored today, the universe is made of matter, not antimatter. What happened to all the antimatter that we expect? At one millisecond of time (one thousandth of a second) after initiation of the Big Bang the environmental conditions favored formation of quarks that make up subatomic particles such as protons and neutrons. Any collision between a quark and anti-quark released an amount of energy equivalent to the sum of the two masses (just as with the electron and positron annihilation). Such collisions happened at a high rate due to the high density of matter at this time and at this location. It turns out that for every billion quarks and anti-quark pairs that annihilated each other there was an extra ordinary matter quark. Only those extra one in a billion quarks make up the mass of our universe today. For every quark in our universe there were one billion quark pairs that did not make it. The release of these quark and anti-quark annihilations must have been huge, even by early universe standards.

Intuitively, a balance is expected, not a slight imbalance in quarks because most basic natural laws imply some sort of balance. For example, conservation of mass-energy (First Law of Thermal Dynamics) states that the mass-energy within an enclosed system stays constant, i.e., stays exactly balanced. Other laws, such as Maxwell's Equations that describe radio waves, microwaves, light, and x-rays are expressed in terms of four exact balances. This balancing appears to be the pervasive pattern in nature, why is there this one part in one billion imbalances for quarks and anti-quarks at the beginning of the Big Bang? The answer to this question is unknown at this time.

However, what we do know is that if there was perfect balance between the quarks and anti-quarks at the time of the Big Bang, there would have been complete annihilation of all matter and there would have been no atomic elements, no biochemistry, no life forms, no life development, no intelligent and self-aware life, no planetary development, no stabilizing moons, no solar systems, no stars, no galaxies, no galactic clusters, and no universe. Despite our lack of understanding on this issue, what we do know is that the universe would not have become structured for conscious life without a one part to one billion parts imbalance between quarks and anti-quarks, which appeared at the beginning of the Big Bang. Without this imbalance, life as we know it would not have evolved.

Next, we turn to a discussion of atoms, their nuclei, and the weak nuclear force, one of the four fundamental forces in our universe.[9] Some atomic nuclei are unstable and lose energy by ejecting particles, called radioactive decay and controlled by the weak force. If the weak force was just slightly weaker most of the hydrogen (an atom with one electron) in the universe would have converted to helium (an atom usually with two electrons), making water very scarce, if it formed at all.

In addition, as is well known, in a universe that is structured or tuned for the development of life, water is one of the most fundamental components. A slightly weaker weak force would have likely translated into an absence of life altogether. If the weak force was slightly stronger little or no helium would be produced from the Big Bang. Without the necessary concentrations of helium, heavier elements (such as carbon, oxygen, and

---

9. Gravity, the strong nuclear force, and electromagnetism are the other three fundamental forces.

iron) would not form in sufficient quantities to support the development of life as we know it because they would not be made by the nuclear furnaces inside stars.

Staying at the size scale of atomic nuclei, we next consider another of the four fundamental forces of nature, the strong force. This is the glue type force that tightly holds together the particles in the nucleus of an atom (protons, neutrons, and others). If the strong nuclear force were just a little weaker by about 2 percent, then all atoms more complicated than hydrogen (atoms with more than one electron) would not be stable. Hydrogen would be the only element in the universe. Furthermore, we know that in addition to hydrogen, life requires other atomic elements and would not be possible if the strong force were altered slightly in the other direction. For example, if the strong force were slightly stronger by as little as 1 percent, there would be a much lower density of hydrogen in the universe. If the increase was 2 percent, atomic nuclei (due to the absence of protons) could not have formed, yielding a universe without atoms. Thus, based on the above scientific knowledge, we are led to only one conclusion: the universe appears to be structured for conscious life.

Randomness and Design

Building on the above conclusion that the universe appears to be structured for conscious life, we now turn to a discussion of the difference between randomness and design. What is clear from the above discussion is that there is an amazing orderliness about the cosmos and that slight variations in the constants that make it what it is would lead to a very different place. Could the patterns that comprise this well ordered world have occurred at random? In this section, we will examine this issue of randomness in depth and contrast it with the concept of design.

We begin with a description of what we mean by the terms design and randomness. We will do this in the following way. Based on Figure 4 as it appears below, we will consider the Mount Rushmore granite carving of the four presidents and compare it to a shadow that appears to be that of President J. F. Kennedy (JFK). We will also look at an example of how random materials might come together randomly and form something of value.

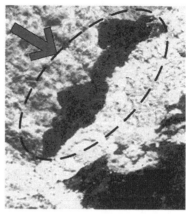

Figure 4. Left: A rational design: Mount Rushmore with presidents Washington, Jefferson, Roosevelt, and Lincoln. Right: An accident of nature: President John F. Kennedy's profile formed by shadow cast by a large rock.

When we look at the Mount Rushmore National Memorial in South Dakota, we recognize the hugely carved sixty feet high sculptures of former US Presidents George Washington, Thomas Jefferson, Abraham Lincoln, and Theodore Roosevelt, which was completed by design in 1941. We also see what appears to be a silhouette of JFK that was formed by a large Hawaiian rock when it cast a shadow at the right time of day.

The JFK shadow differs in a substantial way from the faces on Mount Rushmore. JFK's silhouette is a two dimensional object without thickness. What is of particular interest is that its presence on the ground is not due to any intentional design features. Instead, its appearance has resulted from randomness. The details are imperfect, and the image varies depending on the time of day and the angle from which it is viewed. On the other hand, the features on the faces of the four Presidents never change because they have been carved out of granite by the artists who designed them and the artisans who sculpted them.

Building on this simple but clear comparison, we can elaborate many of the recognizable differences that exist between objects that are created by design and those that appear through randomness. Another example of something highly valued clarifies this distinction further. We start by imagining the existence of a remote junkyard. It is filled with castoff items such as old rusty cars, cinder blocks, random parts of brick walls; construction; plumbing and electrical debris; derelict appliances (such as washers, dryers, and water heaters); nonfunctional auto parts; castoff plumbing parts; and various air conditioner units that do not work anymore.

This debris is piled high in the center of this junkyard. A large explosive is placed under the pile. Everyone moves back to a safe distance, and the explosive is detonated. Parts from the junkyard are randomly torn apart and fly off at high speed in random directions. Many of the parts collide with each other, and all of the parts come to a new resting place. The dust settles, and we move in to see what remains.

There in the middle of what was the junkyard is a new car without blemish. It has a complete voice activated navigation system, memory power seats, leather interior, 12 speaker premium sound system, xenon headlights, air conditioning, six speed shift with automatic transmission, the smell of a new car, and an engine that will take the driver from zero to 60 mph in 7.2 seconds. Every one of the thousands of components of this has been assembled so that the car will operate flawlessly under all conceivable conditions and over 100,000 driving miles.

Normally, the production of this or any other new automobile requires careful design. It must be manufactured with very tight tolerances and meticulously assembled. No matter how long it took, we would rightly say that such a precision designed, functional, and assembled new car could not have originated from such original junkyard materials, an explosion, and undirected processes. The new car could not have self-organized from the original materials in that junk yard. It is virtually impossible for it to happen this way, even if we allow for a time span of 4.5 billion years, which is the age of the Earth.

In light of this example, it is reasonable to conclude that producing such a new car would take thousands of purposeful engineering person hours of investigation and design time, powerful computers to conduct all the required numerical simulations, engineers to develop the software, testing facilities and quality engineers to qualify the designs on test tracks and environmental chambers, integration engineers to successfully merge all the component systems in the final car, vendors to deliver all the necessary components on time and within specifications, manufacturing engineers to design and qualify all the robotics and manufacturing systems, assembly technicians to build the cars to specifications, and transportation to move the finished car to its final location here in our junkyard.

Thus, on the question of whether a new car of any kind and in any circumstances can produce itself at random from start to finish, we can conclude with a very high level of confidence that the reasonable answer is no. It is much more likely that it was created by intentional design than by random forces. Does our universe also reflect a purposeful design like that of Mount Rushmore or a new car? Based on our current scientific knowledge as presented in the next section of this chapter, the answer seems to be yes.

This argument might seem to be another version of the Teleological position, summarized in chapter 1. In the context of this chapter, however, this line of reasoning is more robust than the Teleological position for the following reason: Our discussion of a purposeful design is not a stand-alone argument but it is a combination with other evidences, summarized in this chapter. This blend makes for much stronger argument than a consideration of only the Teleological position by itself. We will see how this works in a later section where we introduce Nineteen Additional Considerations.

The Anthropic Principle and Design

The universe started in an unimaginably huge explosion some 13.7 billion years ago. All the galactic clusters, all the billions of galaxies, all the gas clouds, all the hundreds of billions of stars, all the planetary systems, all the planets, and all life forms, all of it, ultimately originated in that Big Bang. All of our most recent evidence is that all of humanity is made from the same stuff, the same atomic particles, the same subatomic particles, and the same force fields. All that exists in the universe appears to be constructed of a relatively few kinds of particles and in a rather comprehensible manner.

Why is this? To us the reason appears to be that everything in the universe came originally from the same source at the same time—the Big Bang. There were no islands of unintelligibly huge explosions, which were located in different regions of the universe and which occurred at different times independently of each other. A scenario such as this would likely result in different kind of universe. It would seem that we are all connected to the whole universe because of this common origin.

Water

As discussed earlier, let us again consider a liquid with which we are all very familiar, a liquid essential to life: water. We tend to overlook how amazing water is. Because of its structure and unusual property, it expands when cooled from its liquid to its solid state (ice). Very few other materials do this. If it were not so ice would not float; and lakes, oceans and rivers would freeze from the bottom up. Fish would die. If ice sank in water, much of the heat in the water would radiate away, and the Earth would be substantially colder, inhibiting life development as we know it.

The chemical reactions required for life are too slow in a solid such as ice. Complex life would not likely be sustained as a gas where the molecules are only slightly interactive. An environment that better supports organization and development is needed for life with its requirements for energy production, reproduction, and growth. There is a need for a liquid that supports the necessary chemical reactions essential to living systems: a liquid with the properties of water. Not only that but this amazing liquid forms a kind of glue close to and on its surface. Early in the morning when droplets appear on flower pedals, those droplets are due to this glue (surface tension). It is quite likely that surface tension played a key role in the development of life as discussed in the next chapter. The molecules within the cells of living organism need a solvent for the necessary chemical reactions to take place. Anyone who would be the chemical engineer in charge would probably provide something like water for that solvent.

19 Additional Considerations

The example of water is but one of what appears to be the finely structured design features that comprise our amazing universe when viewed from the perspective of the Anthropic Principle. Are there others? If so, what are they and how are they so closely interlaced that any slight deviation would result in an entirely different universe without intelligent and self-aware life in our galaxy? The answer to these questions can be found in the following Table 1, which includes nineteen additional considerations.[10] This table is meant to show only nineteen examples and is by no means complete.

10. Rood and Treffi, *Are We Alone?*
   Anderson, "The Earth as a Planet: Paradigms and Paradoxes," 347–55.
   Barrow and Tipler, *The Anthropic Cosmological Principle*, 510–75.
   Campbell and Taylor, "No Water, No Granite—No Oceans, No Continents," 1061–64.
   Carter, "The Anthropic Principle and Its Implications for Biological Evolution," 352–63.
   Cottrell, *The Remarkable Spaceship Earth.*

Under the heading called "Description," we have included a particular design feature for that row. Under "If A" and "If B" we have summarized the type of design limit. "Result due to A" and "Result due to B" give the anticipated results if limit A is exceeded or if B is exceeded. This table has to do only with effects associated with our galaxy, our solar system, and those associated with our Earth.

**Table 1. Nineteen additional considerations
for the design of intelligent and self-aware life.**

	Description	If A	Result due to A	If B	Result due to B
1	Number of star companions	If more than one	Tidal interactions would disrupt planetary orbits.	If less than one	Not enough heat would be produced for life.
2	Parent star birth date	If more recent	Star would not yet have reached stable burning phase.	If less recent	Stellar system would not yet contain enough heavy elements.
3	Parent star age	If older	Luminosity of star would not be sufficiently stable.	If younger	Luminosity of star would not be sufficiently stable.
4	Parent star distance from center of galaxy	If greater	Not enough heavy elements to make rocky planets (like Earth).	If less	Stellar density and radiation would he too great.

Gale, "The Anthropic Principle," 154–71.

Greenstein, *The Symbiotic Universe: Life and Mind in the Cosmos*, 68–97.

Gubbin, "The Origin of Life: Earth's Lucky Break," 36–102.

Hammond, "The Uniqueness of the Earth's Climate," 245.

Hart, "Habitable Zones about Main Sequence Stars," 351–57.

Hart, "The Evolution of the Atmosphere of the Earth," 23–39.

Owen, et al., "Enhanced CO_2 Greenhouse to Compensate for Reduced Solar Luminosity on Early Earth," 640–41.

Ross, *Genesis One: A Scientific Perspective*, 6–7.

Templeton, "God Reveals Himself in the Astronomical and in the Infinitesimal," 196–98.

Ter Harr, "On the Origin of the Solar System," 267–78.

Toon and Olson, "The Warm Earth," 50–57. Ward, "Comments on the Long-Term Stability of the Earth's Obliquity," 444–48.

	Description	If A	Result due to A	If B	Result due to B
5	Parent star mass	If greater	Luminosity output from the star would not be sufficiently stable.	If less	Range of distances appropriate for life would be too narrow; tidal forces would disrupt the rotational period for a planet of the right distance.
6	Parent star color	If redder	Insufficient photosynthetic response.	If bluer	Insufficient photosynthetic response.
7	Surface gravity	If stronger	Planet's atmosphere would retain huge amounts of ammonia and methane.	If weaker	Planet's atmosphere would lose too much water.
8	Distance from parent star	If farther away	Too cool for a stable water cycle.	If closer	Too warm for a stable water cycle.
9	Thickness of planet's crust	If thicker	Too much oxygen would he transferred from the atmosphere to the crust.	If thinner	Volcanic and tectonic activity would be too great.
10	Rotation period	If longer	Diurnal temperature differences would he too great.	If shorter	Atmospheric wind velocities would he too great.
11	Gravitational interaction with a moon	If greater	Tidal effects on the oceans, atmosphere, and rotational period would he too severe.	If less	Earth's orbital obliquity would change too much causing climatic instabilities.
12	Magnetic field	If stronger	Electromagnetic storms would be too severe.	If weaker	Not enough protection from solar wind particles.
13	Axial tilt	If greater	Surface temperature differences would be too great.	If less	Surface temperature differences would be too great

	Description	If A	Result due to A	If B	Result due to B
14	Albedo (ratio of reflected light to total amount falling on surface)	If greater	Runaway ice age would develop.	If less	Runaway greenhouse effect would develop.
15	Oxygen to nitrogen ratio in atmosphere	If greater	Life functions would proceed too quickly.	If less	Life functions would proceed too slowly.
16	Carbon dioxide and water vapor levels in atmosphere	If greater	Runaway greenhouse effect would develop.	If less	Insufficient greenhouse effect.
17	Ozone level in atmosphere	If greater	Surface temperatures would become too low.	If less	Surface temperatures would he too high; too much UV radiation at surface.
18	Atmospheric electric discharge rate	If greater	Too much fire destruction.	If less	Too little nitrogen fixing in the soil.
19	Seismic activity	If greater	Destruction of too many life-forms.	If less	Nutrients on ocean floors would not be uplifted.

This Table summarizes many of the assumed possible outcomes that might give us a life friendly universe or a life unfriendly universe (represented by these nineteen rows). For each row there are only three possible outcomes: 1) If A . . . , 2) If B . . . , and 3) If neither A nor B (A and B are mutually exclusive). This table shows some of the surprisingly many requirements it takes for a planet like ours to be life friendly and that can sustain life for over 3.85 billion years.

Let us see if it might be possible to integrate all nineteen features into one realistic and defensible conclusion. To do that we need to make assumptions here that are realistic; that translate into meaningful outcomes; and are not biased toward a worldview, such as the universe was caused

a designer or that that universe was caused by randomness and with-
it any direction. We start by assuming that each of the three outcomes is
equally likely for all nineteen rows. This favors only randomness and only
the absence of direction for the cause of the universe (randomness). If we
assumed otherwise, such as the only outcome 2) was more likely than the
other two outcomes for each row, we would be biased for a designed universe (designer). In this case we are assuming randomness so that we have
no reason to favor any one of the three outcomes over the other two.

We know this assumption is not exactly right; each of the three outcomes is equally likely for all nineteen rows. However, it is a good place to
start and can be modified later. We also want to assume that each row has
no significant effect on any of the other rows. What happens when we make
these assumptions, which are very generous for randomness and quite restrictive for design? The result is that the probability for randomness yields
a probability for life in our solar system at one chance in a billion.[11] This
is an incomprehensibly large and unexpected number. Next, let us modify
some of the assumptions and still try to keep them reasonable and easy to
grasp. We relax somewhat our assumption that each row has no significant
effect on any of the other rows. Assuming that our result is reduced by one
half or one chance in 500 million, we are still left with an incomprehensibly
small probability for the randomness explanation.

Next, we relax our assumption that all nineteen rows are included.
Let us say that only ten of the nineteen rows are needed. In this case our
result becomes a probability of one chance in 38,000 for randomness. These
simple simulations, including biasing assumptions toward randomness,
yield an overall result—one chance in 38,000—that overwhelmingly favors
design.

11. If you roll a fair (cubical, six-sided) die, the chances of getting, say, a 3 are 1 in 6
or 1/6. Similarly the chances of getting a 3 on a second fair die are also 1/6. If you rolled
both die together, the chances of getting two 3s (a 3 on each of the two fair die) are (1/6)
x (1/6) or 1/36 or 1 in 36. The two previous 6s multiply together to give 36 for both die
for the independent roll of each die. That is the way random chance works with two
independent events (such as the role of the two dice).

So, in our case of 19 conditions (each one necessary for life and each with an assumed
1 in 3 chance or 1/3) the total probability is (1/3) multiplied 19 times or (1/3) x (1/3) x
(1/3) x (1/3) x (1/3) x (1/3) x (1/3) x (1/3) x (1/3) x (1/3) x (1/3) x (1/3) x (1/3) x (1/3)
x (1/3) x (1/3) x (1/3) x (1/3) x (1/3) or $(1/3)^{19}$ or [1/(1 billion)] or one chance in one
billion. This probability (1 chance in 1 billion) that life develops only due to randomness
is incomprehensively small. Here is an example, think of 1 second compared to 37 years.
That is about one heart beat compared to about half a life time. That is 1 part in 1 billion.

Note that it would not make any significant difference if we were to alter the considerations that appear in Table 1 so that there would be more rows or to reduce the odds from one chance in a million to one chance it a thousand. It looks like whatever reasonable assumptions we make, given the well-established multiplier effect of a combination of all the probabilities (as summarized in footnote 11), we would still come to the same conclusion: The probability that conscious, self-aware life would develop on Earth is infinitesimally small and resorting to randomness as a way to account for how this happened results in a much lower confidence level than for the only other alternative: a design explanation.

Why might this be? According to the Anthropic Principle, the universe is so narrowly and tightly designed that a more plausible explanation of how conscious, self-aware life came into being is that it is the outcome of some kind of design influence. In the last chapter, we will address in detail the issue of whether it is more reasonable to conclude that the universe that is characterized by design elements is the result of intentional design or was created at random.

Conclusion

We conclude with a summary of the important issues that have been discussed throughout this chapter. We started with a discussion of the Anthropic Principle and its connection to the development of conscious, self-aware life based on how the laws of nature have operated since the time of the Big Bang. These laws have operated by a set of constants within a very narrow range. Even slight variations outside the parameters of these constants would have led to life forms other than what we know on Earth.

In the process of considering how conscious, self-aware life came to exist, we considered the following scientific findings and knowledge. We describe the significance of the Standard Model, the four fundamental forces by which the universe operates, and the discovery of antimatter. Despite these understandings, we still do not know the nature of the dark matter or dark energy that comprises 95% of the known universe.

At the same time, we indicated that despite what we do not know, there is much that we do. In addition to the Big Bang and its consequences, we described the nature of gravity and its impact on how the universe developed and is sustained in its current condition. We also analyzed the differences between the concepts of design and randomness and how the

Anthropic Principle implies the possibility of design. Based on our discussion of the nineteen conditions that are necessary for conscious, self-aware life to develop on Earth, we raised the question of whether the infinitesimally small probability that this would happen argues for design influences of some kind in the creation of the universe as compared to randomness.

Does such a highly structured universe as understood in terms of the current evidence and the Anthropic Principle lead to more confidence that it came to be by intentional design or through random forces and processes? We will answer this question extensively in chapter 8. However, first we need to look at more scientific evidence in other areas. In the next chapter, we will consider the subject of how life on Earth evolved out of non-life.

4

From Non-Life to Life

Introduction

waste

A n astounding and mystifying event took place on our planet roughly four billion years ago. Dormant, nonliving matter came alive. It started its internal mechanisms to eat, reproduce, and expel *waist* by using available energy and chemicals. Just before this event all matter on Earth was nonliving. Just after this event some of that matter became alive. Scientists continue to seek comprehensive explanations for this unexpected mystery.

Determining when life originated is a tough problem not yet completely solved. Any successful explanation takes us from the simplicity of a geochemical world of rocks, molecules, and water and transforms those materials to something that is as complex as just about anything we know: life. Portions of an explanation must use a wide range of sciences including astronomy, geology, earth science, physics, and chemistry. The eventual successful explanation must connect each piece so we can clearly understand how life began. The challenge in determining how non-life became life is complicated by the fact that we have limited knowledge of the environmental conditions that existed at the time; of the thermal and chemical environment; of bombardment of the Earth by comets, asteroids, and meteorites; and of the possible pathways for those inanimate molecules to become living. In this chapter, we will summarize some of today's leading theories of the origin of life on Earth.[1] As will become clear, the answer

1. "How did Life Begin?"

to the question of how life started remains unresolved, although many scientists have advanced alternative explanations for how this happened. Their theories have addressed several issues. These include identifying a series of necessary preconditions that make life possible, such as the available chemistry, temperature ranges, and types of solids and liquids. They have also speculated about different types of life forms and the requirements for each as well as the degree of planetary stability that is necessary to sustain them. In no small measure, our summary of the major scientific ideas concerning the origin of life on Earth will contribute to helping us understand our place as humans in the universe.

To start we need a definition of life, which turns out to be more elusive than what we might first think. It is probably more instructive to summarize some of life's essential characteristics as widely accepted.[2]

1. Life reproduces itself. Humans have babies and oak trees have acorns. Life self assembles itself (reproduces) and brings order in the presence of an energy source (usually the Sun). Without an energy source, nature's general tendency is toward disorder or higher levels of entropy. From generation to generation you get pretty much the same life forms. All life on Earth uses the stable molecule DNA as an instruction book (genetic code) to pass along instructions for reproduction to succeeding generations.

2. Life survives on available food and does not require large molecules that are generally not available. Typically, small, simpler molecules are assembled in the living cells into more complex structures.

3. Life needs to survive, grow, and reproduce. To do all this requires an energy source. Photosynthesis is used by plants that exploit the energy source of sunlight to make sugars. Our bodies have cells with mitochondria that convert food energy into ATP, a molecule that provides energy for most of the cells' functions.

4. Life has the ability to evolve over time in response to environmental conditions.

5. Life needs a container such as your skin or cell membrane to distinguish life from its environment.

Life needs only a few elements: carbon, hydrogen, oxygen, nitrogen, sulfur, and phosphorus. Only a few other trace elements are used. The

2. Collins, "Making Life."

element used most is carbon. The variety of carbon-based molecules is much greater than that of any other element. Carbon can form at least 10 million different molecules. Liquid water seems to be a fundamental solvent for life, because it is needed for operation of cells and organisms as a transport medium and solvent for the raw materials and energy. It has a wide temperature range that varies from freezing to boiling, which helps life to sustain itself.

The basic unit of life is a cell. Recent evidence has shown that the cell is a complex structure of chemical mechanisms and structures that help the cell perform its functions. The full story of how all this works in the cell has not yet fully been worked out. Energy is needed to drive the chemistry for the origins of life. But how this energy actually works and what is the source has yet to be determined. The principle (but not the only) source of energy for the origin of life is sunlight. Other sources include heat and chemical energy-the energy of breaking and reforming chemical bonds.

While there is widespread consensus on the elements that are essential to sustain life, the focal question of this chapter remains to be answered: How does a diverse collection of non-living molecules come together and start to function in order to morph into the first living cells? Life is self-regulating and a delicately complex set of nano-machinery of a high order. Today there is no comprehensive explanation that describes how life could have appeared spontaneously undirected. The short answer is that we really do not know. One of today's experts Andrew Knoll[3] speculates that there might have been some kind of molecule that balances as much simplicity as possible with enough complexity that it can reproduce in those primitive conditions. When that happened, lifeless matter, molecules, may have morphed into the earliest primitive life forms. This means that the complex molecular structures and biochemistry we see today are almost certainly unlike beginning life forms. This becomes clear when we reflect on what the Earth and atmosphere were like at the time of life's origins.

Earliest Conditions

The radiation that Earth receives from the Sun includes visible light and invisible radiation damaging to life such as ultraviolet radiation and x-rays. For about the first 500 million years after Earth's formation it was exposed

3. Knoll, *Life on a Young Planet: The First Three Billion Years of Evolution on Earth.*

to x-rays (ionizing radiation) that were about fifty times higher intensity than what we have today.[4] Also at this time the radiant power from the Sun was variable as the Sun had just began to stabilize. Using current understanding of how the Sun and similar stars produce power and confirmed by observations of those stars in their early developmental stages, astronomers can reconstruct the life threatening hostility of the Sun's early radiation.[5] The conclusion from these studies is that it is unlikely that life could have started on Earth before about 3.85 billion years ago.

We also know that the Earth took a severe life-devastating physical pounding especially between 3.5 and 4.6 billion years ago from asteroids, small planets, and other rocky bodies.[6] This is called the late heavy bombardment. All of this (and much more) has been carefully examined in regards to the inhospitable environment for life on Earth. The conclusion is that up to about 3.85 billion years ago the Earth was "sterilized" many dozens of times.[7] For each of these events the following was determined:

1. The Earth's surface substantially melted after each event.

2. No liquid water remained on the Earth for at least three thousand years after each event.

3. No solid rocks remained on the Earth after each event.

4. No life forms or pre-life forms or molecules could have survived after each event.

At this time the Earth was a very violent place, almost certainly incapable of supporting life, let alone providing an environment suitable for life incubation.

Next, we turn to the fossil evidence, which is even more difficult to interpret. Rocks that can be dated from 3.4 to 3.8 billion years ago are exceptionally rare, and they are unevenly distributed around our planet. The

---

4. Rana and Ross, "*Origins of Life*." Walter and Barry, "Pre-and Main-sequence Evolution of Solar Activity," 633–57. Soderblom, et al., "Rotational Studies of Late-Type Stars. VII. M34 (NGC 1039) and the Evolution of Angular Momentum and Activity in Young Solar-Type Stars," 334–40.

5. Rhode, et al., "Rotational Velocities and Radii of Pre-Main-Sequence Stars in the Orion Nebula," 3258–79; Alencar, et al., "The Spectral Variability of the Classical T Tauri Star DR Tauri," 3335–60.

6. Rana and Ross, *Origin of Life*, 82.

7. Wells, et al., "Reseeding of Early Earth by Impacts of Returning Ejecta During the Late Heavy Bombardment," 38–46.

Earth's crust has moved along with the continents as they split up from the original single continent—Pangaea—and migrated to their present positions. Those rocks have been melted, moved about, compressed, buried, and heated. All of this obscures valuable information from that ancient time period we are interested in.

All is not lost, however. The formations that are denser in residues of life from this time period are on coasts and near springs of hot sulfurous water. These facts suggest that the rare pieces of geological evidence of early life that have been discovered do not necessarily represent the first regions where life may have started. Traces of early life have been discovered that date back to 3.4 to 3.5 billion years ago. That life was ancestral bacteria and was already differentiated at this time (filamentous, unicellular, multicellular, and forming dense structures on the coast).

These life forms could not have been the earliest forms of life just after their formation because they were too advanced. Estimates and extrapolations have been made that the earliest start date was about 3.80 billion years ago,[8] some have suggested 3.85 billion years ago. Thus, we have two independent sources of information, the fossil record of life and studies of our solar system, which point to a start time for life of about 3.85 billion years ago. In addition, keep in mind the time scale here. 0.05 billion years is 50 million years ago. Humankind developed the city-states of ancient Greece only 2,500 years ago. 50 million is 20,000 times longer than the time between now and the development of city states. These time scales for Earth's development are incomprehensibly long.

All of this evidence summarized here makes it nothing short of challenging to see the big picture about when life might have begun. Figure 5 summarizes and integrates this discussion. On the vertical axis is plotted the probability that life existed from 0 percent to 100 percent. (This scale is not research grade and has uncertainties.) On the horizontal axis is plotted time into the past, "Years ago" in billions of years. The solid curve illustrates the probability that life existed over the time period of about 3.5 to about 4.0 billion years ago, based on the above summary about the Earth's environment and its suitability to support life, especially early developing life. The dashed curve illustrates the probability that life existed over that time period, based on the fossil record and associated scientifically based estimates.

8. Manning, et al., "Geology and Age of Supracrustal Rocks, Akilia Island, Greenland: New Evidence for a >3.83 Ga Origin of Life," 402–3.

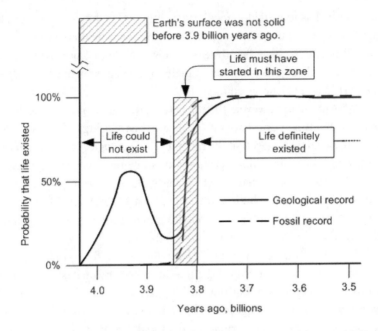

Figure 5. Based on planetary, geological, and fossil records life must have started about 3.80 to 3.85 billion years ago. These two curves are not quantitative.

These two methods (Earth's environment and fossil record) are independent and the fact that they give nearly the same result provides additional credibility to the conclusion. What we see is that based on the solid curve (Earth's environment and geological record) life could not have existed, much less started, earlier than about 3.85 billion years. Based on the dashed curve (fossil record), we can conclude that life must have started by about 3.80 to 3.85 billion years ago. This is a time window of about 0.05 billion years or about 50 million years, which is a rather short period of time when compared to the history of our solar system and the history of the Earth. As we will see below this time window has important consequences for various suggestions for how life started.

Thus, it appears that life began almost immediately after Earth could support it. At that time the Earth was a dark (sunlight did not penetrate the atmosphere at that time), it had no oxygen yet, and it was poisonous and hostile for current life forms that breathe oxygen—but apparently not for those first life forms. One example of hostile environments with life is Mono Lake just east of Yosemite National Park in the high California desert. Its salt density is 2.5 times the salt density of the world's oceans and the salt is a different type. The waters of Mono Lake are alkaline, caustic, and toxic to many life forms. The waters are rich in arsenic. If you drank this water, you would likely die. Yet, amazingly some life thrives in these waters. That life has adapted to its environmental temperature, solvents, and chemical surroundings. If life could adapt to this hostile ecosystem, then that implies that life has the capability to adapt to other hostile situations, such as those found on the Earth 3.85 billion years ago.

Given this background we return to the central question of this chapter: How did life start? As we have already indicated, this is a difficult question to answer because life emerged at a time that was almost entirely unrecorded, even with geological data. This implies that we do not know exactly what those environmental conditions were. In order to completely understand how life started we would need to set up our confirmation experiments in a controlled and realistic manner that is similar to those conditions. The combination of conditions, materials, chemicals, energy, and natural processes that formed life in the beginning was extremely unlikely and statistically nearly impossible. It was one event in a billion billion. But it did happen. Life emerged out of non-life.

Sustaining Life

With this background on earliest conditions, our next step is to look at some of the elements that sustained life after it started. We recognize at the outset that although our knowledge is incomplete the following factors are essential prerequisites without which life would lack long term potential for survival. It would be ideal if we could create experiments in a controlled and realistic manner that duplicated the exact conditions that gave rise to life some 3.85 billion years ago. However we cannot. Instead, our experiments would be statistical in design in order to cover the many components and combinations that realistically supported life.

The initial conditions that we identified in the previous section are complicated because they include temperature ranges and variability, presence of liquid water at the site of life formation, absence of a toxic chemical environment after embryonic life forms, and the presence of molecules and precursors that eventually translate into life. These elements must be stable enough so that the environment does not convert into hostile surroundings. For example, life might first develop in a spot near where a meteorite strikes and disrupts those first life giving molecules. These types of starting conditions must be considered for explanations about sustaining life once it began.

Other considerations must be included as well such as identifying the pathways from the first life forms to complexity and variety of life we have today.[9] Those pathways include the evolution of the initial atmosphere, devoid of oxygen, to the atmosphere we have today, including the oxygen that is so essential for us. Those pathways include changes in the distribution of water—both liquid and ice—and changes in weather. Any explanation that is ultimately successful and based on randomness must accommodate at least the following:

1. Start with the undirected and random assembly of those pre-life molecules and explain the natural influences that translate them into life friendly structures.

2. For the initial conditions include the essential features of the environment when life got its start.

3. Explain how those life friendly structures where changed into the first life.

4. Explain how that first life survived the subsequent changes in its environment.

5. Quantitatively explain the probabilities of how these events occurred.

Ultimately we want to understand how atoms and molecules in that primitive environment randomly assembled to form life that ultimately evolved into our consciousness, allowing us to consider the origins of life and our own destiny. It appears that once life was well underway we have explanations that give us confidence for understanding the subsequent development of the higher life forms. It is just that, so far we have no adequate explanations for how life started.

However, despite our lack of precise knowledge of life's origins, we know that an important type of molecule for sustaining life is protein. Connective tissue, hair, and muscle are made of proteins configured as extended chains. Take another example, hemoglobin, which transports oxygen in your blood. Other types of proteins act as enzymes, which encourage (or discourage) certain chemical reactions that are beneficial (or detrimental) to life. Virtually every chemical reaction in the human body is encouraged by a specific enzyme—a different enzyme for each of the thousands

9. Gribbin, *Alone in the Universe: Why Our Planet Is Unique.*

of reactions. Half of the mass of all the biological material on Earth is in the form of amino acids.[10] A specific protein may consist of hundreds of thousands of separate amino acid units.

This is similar in complexity to a sentence containing every word in English using the twenty-six letters of the alphabet. With hemoglobin, which contains 539 amino acids, the number of different arrangements in which we can place those amino acids is equivalent to a 1 followed by 620 zeroes, an incomprehensibly large number. The huge number of all the sub-atomic particles in the known Universe is essentially zero compared to this number. Yet for hemoglobin to work properly only one arrangement (out of a total of 1 followed by 620 zeroes) is effective. For a protein of hemoglobin to form randomly without any guidance is incomprehensibly small, and a mistake of a single amino acid in hemoglobin can produce a molecule with a dangerous imperfection. There are some limited repair mechanisms in the human cell so that some transcription mistakes can be overcome.

All these enzymes must function within a special environment. All of the components, for example amino acids, must be located close enough to the reaction site for the reaction to occur. Not only that, but the temperature must be within a certain window. The electrical charge distribution on component molecules must be distributed and oriented right. If the component molecule is a long chain of amino acids, the folding of that chain must be amenable to the reaction. These types of requirements (and others) mean that the region where these life reactions occur cannot be in what is called chemical equilibrium. This is where all the constituents, temperature, and other properties are homogenized throughout the space. If those reactants were in chemical equilibrium, it would mean death and life could not be supported there.

In order to support life some type of boundary is required to keep the reactions away from chemical equilibrium. It turns out that all life is organized into cells that provide such boundaries and keep the internal reactions away from the external chemical equilibrium. Have such boundaries been made in the lab? The answer is yes. The oily boundaries—cell containers—are made of lipid molecules that can be shown to spontaneously self-assemble into cell membrane like micro-bubbles called micelles. Some conditions on the early Earth may have been conducive for such cell type containers that might have captured primitive genetic materials. Uncomplicated self-replication of these micro-bubbles may have resulted

10. Hurtley and Pennisi, "Journey to the Center of the Cell," 1399. Rovner, "Bacteria Boast Unexpected Order," 42–46.

from physical pressures. However, we have no clear example where micelles were integrated into sustainable, self-replicating cells.

Beyond the Earth

Is it possible that life originated beyond our Earth, for example on Mars? This concept is called Panspermia, which is the idea that some of the constituents for life were formed in space outside the Earth's biosphere and those constituents seeded the Earth. Scientists have recently discovered that the space beyond our Earth contains certain molecules and amino acids found in the Earth's biology. Some of the same stuff that makes up the cells in the human body is found in the cosmos beyond the Earth. The vast space between the stars of our galaxy does contain water molecules, an apparent essential ingredient for life. Other elements and molecules of methane, certain amino acids, carbon, oxygen, and hydrogen are found in interstellar planet forming dust clouds.

For example, NASA scientists discovered the amino acid glycine in samples of a comet. This was the first time a biologically important amino acid was found in a comet, which supports the theory that some of life's ingredients were formed in space and were delivered to Earth long ago by meteorites and comets. This finding is consistent with the idea that certain fundamental building blocks of life are present in space, and strengthens the argument that certain molecules and amino acids in our biology might be common rather than rare in our universe.[11]

Thus, if the Earth was exposed to the importation of some of life's building blocks, embedded in meteorites and comets, why could not some basic type of life itself actually arrive the same way? If this were the case, then a simple kind of bacteria, for example, could have hitched a ride on a comet to a life friendly environment on the Earth towards the end of the Heavy Bombardment period shortly before life began. While this position provides a straightforward solution to the origin of how life emerged, it is not as simple as it sounds for the following reason. Assume there are biological molecules and materials on or near the surface of Mars, Europa (a moon of Jupiter), Titan (a moon of Saturn), and/or some other location in our solar system. That pre-biological material (PBM) with potential to give the origin of life an assist must take the following types of steps to find its way to a life friendly environment on Earth:

11. "NASA Researchers Make First Discovery of Life's Building Block in Comet."

1. The PBM has to be deposited onto or within a rock on (or near) the alien surface. That PBM has to be attached to the rock so that when it is blasted off the planet's surface (or Moon's surface) the whole package stays together and does not separate. This means, among other requirements, that the rock must have enough structural strength to withstand the blast (or eruption).

2. That rock has to be blasted away from the planet's surface. There are tight requirements for the energy of this blast, which must be in a kind of Goldilocks zone (not too strong, not too weak, but just right). The blast must be strong enough to free the payload (PBM) and rock from the planet's gravitational field. The energy of the blast must be weak enough so that the rock and payload cannot be destroyed so that the subsequent steps can occur. There are several possible energy sources to do this such as volcanic eruption and meteorite impact.

3. Once free of the planet's surface the rock and payload must travel on an interplanetary trajectory that will eventually bring the rock and payload under control of the Earth's gravitational field. This trip might take thousands of years or more. During this long time the payload will be subjected to the dangers of space travel such as high energy cosmic rays, ultra-violet radiation from the Sun, gravitational fields of other planets, and collisions with other interplanetary rocks and meteorites.

   There will be large swings in payload temperature as different sides of the rock face the Sun during this trip. It is possible that embedded radioactivity will supply some warmth to the PBM during its travel but that energy source will decay over time. This warmth must also be in the Goldilocks zone, in order for the molecules of the PBM to disassociate when the temperatures become too high or possibly crystallize when they become too low, which is not as great of a potential problem compared to high temperatures.

4. The rock and payload has to survive entry into the Earth's atmosphere without excessively damaging the payload. During entry the temperature of the payload will increase to approximately that of molten rock. Most meteorites that bombard the Earth do not make it to the surface for this reason. The NASA Space Shuttle is protected with ceramic tiles in order to survive high temperatures upon reentry into the atmosphere. This fourth point and the previous three might have been partially satisfied by the large meteorite that fell near Murchison,

Australia.[12] It contained not only amino acids but thousands of other organic chemicals. The amino acids were found to be enriched in the left-handed isomer and N_{15},[13] indicating that they might have been formed extra-terrestrially.

5. The rock and payload must land on the Earth's surface such that the biomaterials can be unpacked from the interplanetary rock and start to take advantage of the life friendly environment, assuming it landed in such a location. There must be a match between the alien PBM and the life friendly landing site. Once the biomaterials are free enough from the rock they can eventually start to use available energy on the Earth's surface to survive, germinate, and replicate. This critical point was not satisfied by the Murchison meteorite. Each of these five steps must successfully occur for Panspermia to work, that is to say, some of life's constituents were formed in space outside the Earth's biosphere and seeded the Earth.

In addition, using our best current knowledge and the basic laws of biology, chemistry, and physics, we can project the probability that life on Earth began beyond the Earth through a ten step process, although this might involve more than ten. Let us assume the number of required steps is only ten. Let us also assume that the total probability of success for each step is one chance in five considering time spans, space, temperature changes, shock pulses, packing/unpacking of the PBM, and other effects. While this is a generous estimate, the actual probabilities are likely to be much less than one chance in five, perhaps one chance in a hundred or less. Using these conservative estimates, the overall probability that the Panspermia scenario will work for each alien rock with PBM is one chance in 5 X 5 X 5 X 5 X 5 X 5 X 5 X 5 X 5 X 5. This is equivalent to one chance in a ten million. This result would likely be somewhat less for the Murchison meteorite. We see that the chances of Panspermia working are incomprehensively small. Although we might argue that the individual probabilities are not exactly one in five, it really does not matter. What matters is that the result is incomprehensively small. This suggests that our time might be better served by examining other possible theories about the origins of life on Earth.

---

12. "Meteorite That Fell in 1969 Still Revealing Secrets of the Early Solar System."

13. These "left-handed" amino acids cannot be superimposed on the mirror image of the right-handed version of that molecule. It is not possible for all the major characteristics of both hands to match.

However, even if Panspermia is the right answer, it does not answer the question of the origin of life and its constituents on Earth. It merely changes the origin to somewhere else beyond the Earth. For Panspermia to be a believable explanation there need to be believable reasons based on our knowledge of biology, chemistry, and physics for each of the steps from inanimate molecules to early life forms. Starting with accepted science, each step needs to be identified, worked out in detail, and simulated taking into account appropriate probabilities. When all is said and done, we conclude that at this time the Panspermia theory has not provided adequate scientific answers for the question of how life originated on Earth.

Can these inadequacies be overcome? The answer is probably no for the following reason. Even if we assume that the incomprehensibly small probabilities that we identified above can be overcome, we would be left with the possibility that there exist one or more places in our solar system beyond the Earth where PBM or primitive life itself could originate in the sense that those alien environments do not exclude PBM formation. Using the same steps, there is nothing in our assumptions that would exclude the possibility that life could travel in the other direction. Just as life on the Earth might come from alien surfaces, so might life on alien surfaces come from the Earth. Simulations have shown that if Earth rocks undergo collision events that propel them beyond the Earth's gravitational field, 29 percent would land on Venus, 2.0 percent on Mercury, 1.7 percent on Mars, and smaller percentages would travel on to Jupiter and Saturn and their moons.[14]

Thus, we conclude: just because we might discover PBM or even primitive life on those alien places in our solar system we cannot unambiguously assume that life originated on that body and not on the Earth. Therefore, we are compelled to look for other explanations besides Panspermia to explain the origin of life on Earth.

Primordial Soup

In 1952 a graduate student Stanley Miller, in the lab of his Nobel laureate supervisor Harold Urey assembled some glass containers and connected them with rubber tubes. He wanted to simulate the early atmosphere and oceans about the time life got started on the Earth, a kind of "primordial

---

14. Rana and Ross, *"Origins of Life,"* 203.

soup." Into one container he placed water and into the other he placed a mixture of gases, including methane, ammonia, and hydrogen sulphide. He added some heat to imitate the Earth's environment at that time. Atmospheric lightning was approximated with electrical sparks within this apparatus. Then he turned it on and allowed it to run for one week after which the water turned green and yellow.

Upon analysis it was shown that the colored water contained two of the simplest amino acids (glycine and alanine) and other organic compounds. This seemed to show that life emerged out of non-life precursors when favored chemical reaction assembled important biological compounds. Miller believed that only a little more work was needed to provide a complete explanation of the origin of life. The problem would be solved and one more mystery would yield to human wisdom and rational inquiry. The press coverage was extensive.[15] Despite the fact that the gap between a few amino acids and the fragile RNA molecule needed for replication is a very big chasm, Harold Urey was quoted as saying, "If God didn't do it in this way, he missed a good bet."

It turned out to be much more complicated than either Miller or Urey initially imagined it to be. Since those first simulations in 1952, not much progress has been made even though synthesizing amino acids does not appear to be too difficult of a problem. When life began, some amino acids were perhaps produced under conditions mimicked in Miller's experiment. Some might have arrived on Earth's surface in meteorites and comets, perhaps scavenged from interstellar gas clouds like those where amino acids have been observed. It appears possible that amino acids were formed in the Earth's early "soup."

The real problem focuses on proteins and the protein basis of life. As explained earlier in this chapter, proteins are required for life. They consist of chains of amino acids that must be in the right sequence. Also proteins must have the exact folding and three-dimensional structure to function correctly. For example, mad-cow disease and other diseases are caused by unfolded and mis-folded proteins called prions. It is for these reasons that many scientists currently working in this field assume that life started with simpler protein type molecules; and in response to changing evolutionary pressures those molecules evolved to have more complexity.

15. Horgan, *The End of Science*, 138–142.

DNA is the molecule that carries genetic information, used for cellular replication.[16] A simpler and related molecule found in the cell is RNA that includes genetic information and is used to form proteins in the cell. This implies that the earliest life forms were simpler forms of RNA that were able to copy themselves without enzymes and other proteins. For example, some simple membranes of fatty acids have been shown to have characteristics of replication.[17] Random mutations might have kicked-off the initial self-replication processes. Since this would have been the first form of life, there were no predators to worry about. Some cells prospered more than others because they were better at replication, better at using the available energy sources to drive their internal chemistry, better at surviving the challenging early Earth environment, better at expelling the waste products within the cell, and better at adapting to changes in their environments. As these flourishing but primitive cells continued some of them mutated to variations based on their simple replication methods. Most of those mutations failed and those cells went extinct.

However, a few mutations resulted in improvements in the reproductive processes. After millions of such successful mutations, the first primitive RNA type molecule appeared. It might have taken 100 million years or so for these reproductive changes to occur. We can see that once life starts then these types of evolutionary processes take over. More and more efficient changes favor the surviving cells while those mutations that are not so efficient produce cells that tend to die out.

Recently researchers have found an evolutionary pathway for chemicals, present on the early Earth, to mutate into a primitive RNA type molecule. Since DNA is more stable than RNA, DNA might have appeared after RNA. However, this pathway of chemical progress is highly unlikely when certain naturalistic issues, applicable to the early Earth are considered. These include chemical instability, environmental temperature swings, environmental variations in acidic levels, volcanic activity, impacts from comets and meteorites, and variations in the various energy sources such as volcanic heat, lightning, and sunshine. In addition, given the probable presence of many billions of cells, about 100 million years, and large areas of predator-free space, it is possible that there could have occurred mutations that resulted in reproductive advantages.

16. An excellent explanation of DNA is found in Collins, *The Language of God*, 100–107.

17. Ricardo and Szostak, "Life on Earth," 54–61.

At the same time, this might not be the way that life started and evolved to what we have now for the following reason. Since we do not have a detailed grasp of the Earth's early conditions and environment, we do not know the precise conditions that contributed to the emergence of life. This means that we are still searching for an acceptable scientific explanation for how non-life became life. However, recent discoveries have fueled further speculation on how life might have started on Earth.

Higher Temperatures

Recently hot and mineral rich water flows were found on the ocean floor, called hydrothermal vents.[18] See Figure 6. The combination of heat energy and certain pre-biotic minerals associated with volcanic action produces and sustains the energy source for these vents. Unheated seawater seeps through openings in the seafloor where the water contacts a region that has been heated by molten rock and seeded with minerals released from those rocks. The thermal energy inside the Earth is conducted close to the Earth's surface at tectonic plate boundaries or hot spots where that energy raises the temperature of the molten rock. A similar effect with the same source of minerals and warm waters is found at Yellowstone National Park where there are pools of water and geysers that are heated and sustained by the same method.

These vents are interesting because in these harsh, high temperature, high pressure, and mineral rich waters a variety of life has adapted and is sustained at ocean depths where there is a very small density of life because of no light, low oxygen and low temperatures. These life forms at the bottom of the food chain include bacteria that are consumed by tubeworms, shrimp, and even crabs that live in regions around these vents. The primary energy source for the bacteria and ultimately for other life higher up the food chain depends on the minerals and chemicals that are found there and supported to a lesser extent by the heat energy from the hydrothermal vents. The vent ecosystem contains a very wide range of temperatures, mineral concentrations, and rich thermal and chemical energy sources. New ecosystems have been discovered that depend on these energy and chemical sources from the vents.

18. Howland, *The Surprising Arcaea: Discovering Another Domain of Life*," 19–48. See also "Vents and Volcanoes,"

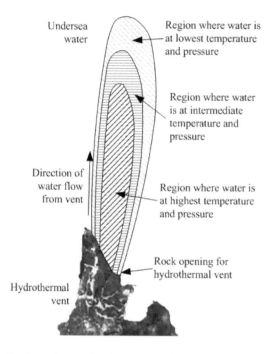

Figure 6. A hydrothermal vent, showing associated regions of water temperature and pressure. The density of minerals approximately follows temperature and pressure.

About the time life appeared on the early Earth, approximately 3.85 billion years ago, there was an abundance of minerals and chemicals, similarly supported by heat energy from impacts from meteorites and comets and from volcanoes. Minerals and chemicals were being released from rocks associated with those impacts. Today there are many examples of diverse life forms near hydrothermal vents. Could life have also started near one of those ancient hot spots? Maybe the first life was a simple, sulfur-loving bacterium living in the presence of thermal energy. In these places there certainly were sufficient minerals. Actually this type of bacterium is part of a larger class of life forms that live in several kinds of extreme environments. See Table 2 below.

Might life have started as a simple version of one of these extremophiles? Even though this might be possible, the answer is probably no for the following reason. Just because life is adaptable enough to exist in these extreme conditions now does not mean that it started in the same conditions. Furthermore, there is no fossil or other scientific evidence that links the origin of life to subsea vents.

Two different scenarios help make this point clear. In scenario one, we can imagine that there are pre-biotic materials (PBM) in a mineral and chemical rich, ancient, and warm environment on the Earth about 3.85 billion years ago. The PBM self-assemble by a yet to be determined process into the earliest, simple life single cell forms that are heat loving and have the characteristics of life mentioned in the Introduction of this chapter. The second scenario takes place in modern times. An existing single cell bacterium moves from a benign environment but with predators into an extreme environment such as a mineral rich, high temperature region near a hydrothermal vent and without predators. This new environment has the disadvantage of higher density of heat energy but it also has the advantage of no predators in the area.

Now let us assume that the absence of predators is more important than the discomfort of high temperature. Survival is a very strong motivator in all of life, usually even more than discomfort. Furthermore, we can assume that the bacteria are able to remain and survive in this predator-free zone for generations. Because of the well understood effects of genetic mutations that cause changes in our bacteria from one generation to the next, they evolve into higher mineral concentration and temperature loving single cells as time goes on from generation to generation. As an example a convincing, long term study and explanation of these types of evolutionary effects on bacteria can be found.[19]

19. Dawkins, *The Greatest Show on Earth*, 116–33.

Table 2. Examples of life that lives in extreme environments.[20]

Name of extremophile	Type of extreme environment	Comments
Thermophile	High temperature	Typical temperatures are 120 - 160 degrees F (50 - 70 degrees C). Found in hot springs and undersea vents.
Hyperthermo-philes	Extremely high temperature	Typical temperatures are 176–235 degrees F (80–113 degrees C).
Psychrophiles	Cold temperature	Typical temperatures are about 39 degrees F (4 degrees C). Found in the Arctic Ocean.
Acidophiles	Acid environment.	pH is less than 2 (neutral pH is 7).
Alkalophiles	Alkaline conditions	pH is greater than 10.
Halophiles	Salt mines & salt lakes	These environments contain from 20 to 30 percent salt.
Barophiles	High pressures.	Locations include Pacific Ocean's Mariana Trench (the deepest sea-floor depression in the world). Barophiles grow best at 300 to 700 times sea-level air pressure.

Notice that both scenarios produce mineral loving single cell life forms in the presence of heat. Yet the causes and origins were fundamentally different and unrelated in each scenario for those single cells to be well adapted to their extreme environment. This same general argument can be applied to each of the extreme environments given in Table 3 below. The conclusion is that we cannot extrapolate backward in time and accurately determine the causes and origins of life formation, starting with current observations of extremophiles. Just because thermophiles exist today around hydrothermal vents does not mean that life started that way.

The explanation for the origin of life must stand on its own. The process for a successful explanation would include the following steps. We would need to start with the conditions on Earth 3.85 billion years ago. Then we would have to use known physics, chemistry, biology, and probability

20. Rana and Ross, *Origins of Life*, 172.

theory; compare the candidate explanation with competing explanations in detail; and compare the predictions of the competing explanations with known, relevant, and verifiable observations. The explanation that best matches known scientific facts would prevail as superior to the others until such time as a more successful explanation comes along.

The Role of Clay and Minerals

Could clay have played a role in the formation of life? The surfaces of some clays have properties that attract biological molecules in a liquid solution, such as a primordial soup discussed above. Those molecules tend to be concentrated and the clay's surface provides a substrate on which biological molecules have a higher chance to react. The clay surfaces tend to hold and orient molecules so that the active sites have a higher chance of coming in contact and the reactions can proceed faster and with more efficiency. For example, clay surfaces have been shown to be catalyst for the building up of RNA molecules.[21] Recall that RNA, which is similar to DNA, is involved in genetic processes. This implies that the evolutionary progress, based on chemical mutations, can be more efficient from cellular generation to generation.[22] The current conclusion of many scientists is that clay surfaces may have facilitated early biological reactions, but that it is not a complete explanation for the origin of life. While there are other explanations for the origin of life, due to space limitations we will not discuss them here.

21. *How Life Began.*
22. Graham, "Clays Could Have Encouraged the First Cells to Form."

Conclusion

Table 3. Examples of some factors necessary to explain the origin of life.

Factor	Known uncertainties
Atmosphere	Temperature, pressure, chemical components, amount of ultra violet protection.
Water	Amount of water vapor, temperature, Solutions dissolved in liquid water and their concentrations.
Impacts due to meteorites and comets	Frequency, force, and location of impacts & their variations; amount of sterilization from each impact.
Sun	Energy level incident on the Earth and its variation. Energy distribution on the Earth. Reflection of energy back into space.
Earth	Amount, distribution, variability, and strength of volcanism. Amount and rate of planetary cooling.

Based on the above discussion, we can draw several conclusions. To begin, it is clear that there are several reasons why explaining the origin of life is so difficult. The first is that it happened in the extremely far distant past, 3.85 million years ago. Using all our knowledge gained from space probes and telescopes; our understanding of the development of the Sun, solar system, and of the Earth; and our understanding of the applicable science, we are not able to correctly extrapolate that far back and with enough accuracy to predict the conditions on the Earth's surface at that time. Simply stated, we do not have accurate scientific knowledge about the starting conditions. (See Table 3) Although it is possible that life on Earth might have started beyond the Earth, the chance that it did is quite remote. Even if this were so, the origin question would only be shifted to another celestial body where further speculation is nearly impossible. Furthermore, we cannot duplicate those conditions in the laboratory in order to start a set of experiments.

Nonetheless, despite our lack of current knowledge, we could speculate about the conditions that might have existed at the time of this long-ago event. For a satisfactory explanation for the origin of life that is likely to be well received in the scientific community, the following conditions (or similar) would need to be met.[23]

1. The required chemical reactions and the required precursor molecules would need to be plentiful on the Earth at the time life originated.

2. Those molecules would need to have the right range of concentrations.

3. Subsequent chemical reactions would depend on initial chemical reactions where the products of those reactions would remain stable and concentrated enough for subsequent chemical reactions to occur.

4. Chemical interference in the origin of life would not occur.

Perhaps one day scientists will arrive at a consensus on the precise nature of these early conditions. However, until such time as this occurs, our overall answer to the question of how life on Earth emerged out of non-life is that we do not know.

23. Rana and Ross, *Origins of Life*, 111.

5

The Brain Is Structured
for Spiritual Experiences

Introduction

In this chapter we move from the macro-issues of the universe and the origins of life on Earth to a micro-focus on the human brain. While scientific research in this area has progressed steadily during the past half century, our understanding the structure and functions of the brain is still far from complete. It could hardly be otherwise given that there are about 100 billion neurons or brain cells, each one with one hundred or more connections with other neurons and other types of cells.

At the same time, there has emerged a collection of observations and measurements that form the central thrust of this chapter, which is to examine the connection between the brain and human spirituality. The studies that we summarize include a size scale all the way from the molecular and cellular levels to the whole brain itself. As we progress through the chapter, based on the currently available and relevant evidence, our confidence is somewhat strengthened that the human brain is structured for spiritual experiences.

Our Astonishing Brain

We begin by highlighting some of the most amazing facts about the human brain.[1]

- An adult brain weighs about three pounds. If a person's body weight is 150 pounds, the brain is only 2 percent of the entire body by weight.

- Brain weight is not the whole story. Albert Einstein's brain was only 2.7 pounds.

- The average brain is comprised of approximately 100 billion neurons (brain cells). For all intents and purposes, this number is next to impossible for most people to comprehend. One way to grasp this extraordinary complexity is to compare it to the billions of stars in our galaxy, the Milky Way, which is roughly the number of neurons in the human brain.

- Mental activity comes at a price: The brain insatiably uses 20 percent of the total energy we burn and the oxygen we breathe. This is approximately ten times the energy and oxygen usage per pound consumed by the brain compared to other parts of the body.

- Brain cells will begin to die after about four minutes without oxygen.

- More electrical impulses are generated in one day by a functioning human brain than by all the telephones that existed in the world in 2009.

- Neurons send signals at speeds of up to 200 miles per hour. Compare this to major league baseball batters who must decide whether or not to swing a bat and, if they do, where and when to swing it in less than half a second for a ninety miles per hour fastball. Successful batters must demonstrate very impressive signal speeds in the brain and nervous system.

- The brain does not experience pain, which means that neurosurgeons can actively touch and prod the brain while the patient is alert.

- The human brain can distinguish about 10,000 different smells and with experience associate many smells with various types of food, specific people, and species of flowers.

1. Wait, "No More Brain Drain: Proven Ways to Maintain Your Mind & Memories."

- A newborn baby's brain is less than a pound. By age five, a child's brain is nearly the same size as an adult's brain (three pounds) and has developed hundreds of millions of new neural connections. Early childhood development, learning, and experiences are clearly important.

- The human brain generates about 70,000 separate thoughts in an average day.

How is all this astonishing performance possible? While we cannot describe here all of the available neuroscience to answer this question, we can summarize the key features that make the brain one of the largest and most complex organs in the human body.

The brain is made up of many specialized areas that work together. The cerebellum is at the base and the back of the brain and is responsible for coordination, stability, and balance. The brain stem is between the spinal cord and the rest of the brain. See Figure 7.

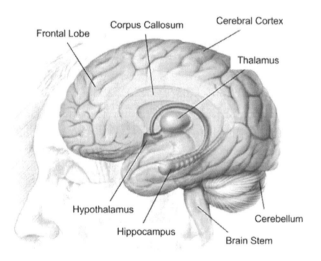

Figure 7. Image of the brain.

Involuntary functions like breathing, heartbeat, and sleep are controlled here. The folded outer layer of brain cells is the cerebral cortex and contains about 85 percent of all brain tissue. Rational thought, perception, motion, planning, and voluntary movements begin in the cerebral cortex. The basal ganglia are a cluster of structures in the center of the brain, which coordinate messages between other brain areas.

The brain is also divided into several lobes. The parietal lobes are responsible for body positions, sensations, perceptions, and handwriting. The frontal lobes are responsible for judgment, problem solving, and motor functions; for example, using mostly his frontal lobe, a major league baseball player decides whether or not to swing at a ninety miles per hour fast ball (145 kilometers per hour or 40.2 meters per second). Hearing and memory are handled by the temporal lobes. Visual processing is done in the occipital lobes.

Previously our general understanding was that the brain is an organ that slowly matures from birth to young adulthood after which it slowly decays as we age. Current understanding, on the other hand, is that our brains are under constant structural change. Tentacle type receptors, dendrites, extend from brain cells (neurons). Dendrites can grow and retreat over several weeks, sometimes as short as a few hours. What this means is that due to a neuron's environmental change, different information can be sent to different parts of the brain with various signal strengths.[2] It must be concluded that neurons are constantly changing, caused by competition, environmental changes, and learning.[3]

The current estimate for the total number of genes that comprise our body's total genome length is about 23,000. At least half of these are dedicated to the assembly, adaptations, and maintenance of our brains. This indicates that in addition all the other thousands of functions of genes, the brain and associated signals to and from the brain are determined to a certain extent by genetic factors.

The Newer Part of the Brain

The recent neurological evolution of the brain suggests that empathy and social awareness are the most recently developed parts of this organ. For the most part, our brain was adapted to survive in an environment that in our first 120,000 years of development was incredibly harsh. This is approximately when *Homo-sapiens* first appeared in East Africa to the present

2. Gaiarsa, et al., "Long-term plasticity at GAB Aergic and Glycinergis Synapses: Mechanisms and Functional Significance," 564–70.

3. Polley, et al., "Naturalistic Experience Transforms Sensory Maps in the Adult Cortex of Caged Animals," 67–71. Also see Frostig, "Functional Organization and Plasticity in the Adult Rat Barrel Cortex: Moving out-of-the-box," 1–6; and Newberg and Waldman, *How God Changes Your Brain*, 14.

time. During this interval our brain managed to make it through those millennia without the benefits of modern medicine, indoor plumbing, clean drinking water, or cellular phones.

We lived in small groups that competed for limited amounts of property, food, and wealth. Two opposing influences were at work, fueled by the development of the language centers situated in the frontal lobe. The old reptilian part of our brain selfishly struggled for survival, while newer, more fragile parts of the brain strove to form cooperative alliances with other people in our groupings. There is evidence for this in the structure of our brains.[4]

We hypothesize that as the newer part of the brain (anterior cingulate and frontal lobes), N, became more influential, the older reptilian part of the brain (limbic system), R, became less influential on human behavior during human evolutionary development. Of course, the relative influence of N and R in any one person depends on other factors besides evolutionary influences, such as experiences in early childhood. Thus, there is a scatter about the hypothesized evolutionary influence as illustrated in Figure 8.[5]

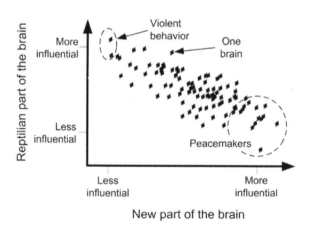

Figure 8. Hypothetical relationship between the new and reptilian (older) part of the brain.

4. Newberg and Waldman, *How God Changes Your Brain*.

5. Figure 8 is given only to illustrate our hypothesized relationship between N and R and not to be interpreted quantitatively.

The general evolutionary pathway is almost certainly from top-left to bottom-right.[6] Also, we can guess that people with a well-developed sense of empathy, social awareness, and peace-making skills appear in the lower right side of the chart with a more influential N and a less influential R. If this is accurate, then it is but a small step to conjecture that people with antisocial tendencies, such as criminals and sociopaths, would probably appear in the upper left of this chart with a more influential R and a less influential N.

Next, we consider the overwhelming evolutionary pressures on our ancestors just after anatomically modern humans arrived on the scene in East Africa about 200,000 years ago. At that time environmental pressures made survival very difficult; undoubtedly, a high percentage of our predecessors perished in those extremely harsh conditions where predators occupied the same lands. In addition, the climate was changing, which added even more survival stress on our early ancestors.

The ancient reptilian part of the human brain responded in a way that was similar to nearly all other animals on the planet. That response was in a selfish, angry, and violent fight for survival. However, the newer, more fragile part of the brain began to emerge, which resulted in the development of language, spiritual sensitivity, self-awareness, and cooperation with others. The brain began to perceive its own existence; and as a result, we humans started to cultivate an internal sense of self-awareness.

The old reptilian part of our brain was beginning to conflict with the newly emerging fragile part of our brain as Table 4 shows.

Through the process of evolution, embryonic religious traditions began to emerge along with their specific concepts of the divine, ways to worship, and methods to meditate. In turn, these emerging traditions strengthened those neurological circuits that favored cooperation, empathy, compassion, and social awareness.[7] Due to neuroplasticity, religious and spiritual meditation started changing the brain in an acute way. In particular, these changes increased social awareness and empathy while restraining the reptilian part of the brain.[8]

6. Newberg and Waldman, *How God Changes Your Brain*, 17–18.

7. Newberg and Waldman, *How God Changes Your Brain*, 17–18.

8. Newberg and Waldman, *How God Changes Your Brain*, 14.

Table 4. Comparison table for older and newer parts of the brain.

Some characteristics of the older reptilian part of the brain	Some characteristics of the newer part of the brain
Self-preservation	Help those less fortunate
Survival	Share the food equally
Anger	Patience
Cheat	Kindness
Take any woman (or man) you want	Generosity
Eat until satisfied, never mind the others	Self-control
Protect your progeny	Follow-through on your promises
Selfish	Don't cheat

If this were not the case, then there would need to be substantial evidence for how other environmental pressures would produce such a newer and fragile part of the brain opposite to the older part. At this point, hard evidence does not exist for explaining the evolutionary conditions that gave rise to the newer part of the brain, especially given the conflicting conquest imperatives that drove the reptilian part of the brain as the above table shows.

Nonetheless, what is clear is that the emerging sense of spirituality was tied closely to the tendencies toward cooperation, empathy, and compassion. Thus, even though we lack reliable scientific knowledge on the specific details of this evolutionary process, what seems to be clear is this: 1) as the brain developed over time, it stimulated the emergence of traits that we identify with the newer and not the older part of the brain; and 2) that these traits evolved simultaneously with the emergence of religion. In a word, the brain became progressively structured for religion and spirituality.

If this conjecture is correct, then there should be scientific studies that confirm that the brain is closely coupled with human spirituality. Does such foundation of evidence exist? The answer is yes as will be demonstrated throughout the rest of this chapter. In particular, our examination of this evidence will focus on the following five areas: God concepts in children, twin studies, near death experiences, the God gene, and the persistence of commitment to religious beliefs. However, before we turn to discussing these five research topics and in order to avoid later confusion, we need to clarify the difference between nature and nurture.

Nature and Nurture

At the most basic level there are two major factors that influence our behavior. Since our behavior is determined by our brain, these two major factors influence our brain as well. They are nature and nurture. Nature refers to the genetic code with which we all start life as newborns, the DNA. This code remains with us throughout our lifetime. Nurture refers to the environmental influences on our brain, such as our families, friends, childhood experiences, and schooling. There is much disagreement on the relative importance of nature and nurture and how these two factors interact with each other.

From our summaries earlier in this chapter we understand that as newborns our brains are structured according to a genetic code given to us by our parents, which supports a nature influence. We also know that our brains are constantly changing and morphing into new structures in response to what happens to us and what happens around us, which supports a nurture influence. In our discussion of the five following arenas of research, starting with concepts of God in children, we will examine the relative influence of nature and nurture factors that cause children to develop their spiritual perceptions and sensitivities.

Concepts of God in children

Recently pictures were analyzed that were made of God by children with different ages.[9] Children from ages six to sixteen were asked to make images of God. There were two matched groups except for one characteristic. In one group the children were raised in a religious environment and in the other group the children raised in a non-religious (irreligious) environment. These children from ages six to sixteen were asked to artistically make images of their idea of God. Evaluative criteria were established to determine whether or not a given image of God was abstract or realistic. All of the resulting artwork was evaluated using these criteria. The results were consistent in four different studies.

9. Newberg and Walman, *How God Changes Your Brain*, 87. Also see Harms, "The development of religious experience in children," 50; Hanisch, *The graphic development of the God picture with children and young people.* Hanisch, "Children's and Young People's Drawings of God."

This gives more strength to these conclusions: Children who were raised in a religious environment clearly developed abstract concepts of God earlier than children who were raised in an irreligious environment. See Figure 9 where the vertical axis is percentage of the children whose artwork showed God abstractly and the horizontal axis is the age in years of the children. The black diamonds and squares refer to the actual results of the studies and the associated smooth curves refer to the accompanying smooth trends. For example, at twelve years old, according to the trend curves, the irreligious child has an expressed image of God that is abstract 8 percent of the time while the religious child expresses God abstractly 41 percent of the time.[10]

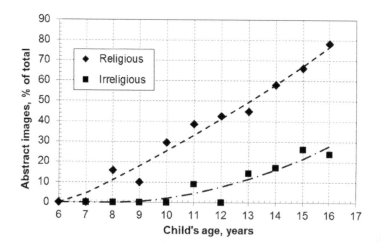

Figure 9. Abstract representation of God by children of different ages.

These studies support the conclusion that the brain of a child (nature) is structured to develop conceptions of God as the child grows (nurture). These studies also reveal that children begin to cultivate their spiritual conceptions early in life. Furthermore, different children develop abstract representations of God at varying levels depending of the diverse environments in which they are raised, such as in families, cultural surroundings, formal schooling, informal groups, and other social settings. Thus, we can conclude that both nature and nurture play a role in how children conceptualize God early in life. By implication, the combination of genetic and

10. Figure 9 was modified from Hanisch, *The Graphic Development of the God Picture with Children and Young People*,"11, and the trends were added.

environmental influences over a person's spiritual development that starts in childhood continues throughout adulthood as well.

Twin Studies

Next, we turn to an examination of twin studies in order to understand more fully how nature and nurture interact in paired adults who started life with the same genetic code. We will look closely at the impact that nurturing influences exerted during their progression through life. Of special significance are studies that focus on twins who possess the same genetic code but who were separated from each other as newborns.

While they are rare, fortunately, such studies do exist. Scientists use two effective methods to search for clues through which to determine the influence of genes on our brains and the resulting behavior. Given the identical genetic backgrounds of twins who were separated at birth, it is easier to identify the impact that being raised in different social environment would have on their development. In addition, twins who were separated at birth are grouped into two types, fraternal (dizygotic or DZ) who do not share the same genetic code and identical twins (monozygotic or MZ) who do. Each MZ has nearly the same genes as the other twin. Each DZ has about half of the genes identical with the other twin.

In one study, researchers at the University of Minnesota compared thirty-one pairs of DZ twins and fifty-three pairs of MZ twins according to several traits. Identical twins shared at birth the same genetic heritage because they possessed matching DNA, whereas the DNA of the fraternal twins did not. The goal of the research project was to identify and contrast the traits that the twins held in common along with those where they differed. All sets of twins were raised apart before the scientist brought them back together in adulthood in order to study their behaviors.[11] By comparing and contrasting the difference traits of the reunited twins, the researchers hoped to determine those that they could identify with genetic origins and others that were caused by environmental influences. They designed their study to determine whether genes would dominate over the environment or vice versa, which is to say, nature versus nurture.

After a careful analysis of their findings, they identified a surprising number of traits that the identical twins had in common, which suggests

11. Segal, *Born Together Reared Apart*, 416.

that genetic influences take priority over those that originate in the social environment. Some of the common traits include the fear of heights, the habit of nail biting, and flushing the toilet both before and after using it. In addition, the MZ twins showed a strong overlap with regard to their spiritual feelings and religious values. Specifically, they were about twice as likely to share a common level of spiritual beliefs, compared to the DZ twin pairs, indicating strong genetic origins for the level of spiritual beliefs.

However when it came to the actual practice of these beliefs, the twins showed few similarities because they held different religious views. This indicates that there is a weak genetic origin for the practice of specific spiritual beliefs, which are the result of environmental influences that stem from the family and other forms of social nurturing. While it is not possible at this point in time to determine where genetic factors leave off and environmental factor begin, it is clear that both nature and nurture play a significant role.

This conclusion is reinforced further in another study conducted at Virginia Commonwealth University on thirty thousand sets of twins—the most ambitious twin study to date.[12] Researchers concluded that "although the transmission of religiousness has been assumed to be purely cultural, studies have demonstrated that genetic factors play an important role in the individual differences in some religious traits." This same team of researchers went on to say that whereas "religious affiliation is primarily a culturally transmitted phenomenon, religious attitudes and practices are moderately influenced by genetic factors." Thus, the findings of the Virginia Commonwealth University research on twins are consistent with those of the University of Minnesota studies, as described earlier in this section.

As scientists continue to search for ways to explain the relationship between behavior and the human genome, perhaps over time, we will better be able to determine how particular genes are connected to specific behaviors, especially those that give rise to human spirituality. Based on the above twin studies as well as those related to conceptions of God in children, we can conclude that there are genetic pre-birth influences on religious attitudes and practices, although it is not clear how this operates at the molecular level within and between the 100 billion (or so) neurons.

12. Kluger, "Is God in Our Genes?" 62–64.

Persistence of Religion

The above summary of research on children and twins regarding the relationship between the brain and religiosity raises other issues that move us to the next step. Clearly, there is a close connection between humanity's genetic predisposition for spirituality and the development of specific beliefs that are connected to the environment in which an individual is nurtured. The next level involves the degree of endurance, or as we might call it—staying power—of spiritual beliefs throughout the course of a person's lifetime. Once they are implanted in the mind, do they persist over time?

Since the rise of modern science, many writers have predicted that religious beliefs that are based on any forms of divine creation or intervention into the laws of nature would give way to purely naturalistic explanations. For example, the quotable Voltaire (1694–1778), famous French Enlightenment figure who influenced important thinkers of both the American and French Revolutions, thought that religion would vanish from the Western world by about 1810. Similar forecasts of this type have frequently been made since Voltaire's time and have come to be known as the secularism or secularization thesis: Religion must decline until it disappears. Modern science will push it into extinction.[13]

In contrast to Voltaire and other supporters of the prediction that religion will one day fade away, 200 years after Voltaire there is strong evidence that the opposite is true. Religious beliefs persist among a majority of the world's populous; furthermore, religion and science can and do coexist compatibly in the modern world. For example, in 2009 in the U.S., the world's most advanced scientific society the Harris polling organization conducted a survey using a representative sample of the American public to identify the extent of their religious beliefs.[14] Pollsters asked 2,309 adults about their religious beliefs and gave them the following instructions:

13. Some authors equate secularism and secularization while others differentiate between them. Secularism refers to reducing all cause and effect relationships to naturalistic explanations and to eliminating all references to supernaturalism or divine intervention. When secularization is defined in this way, it is identical to secularism, which is a theory of naturalistic reductionism. However, secularization can also carry a different meaning and can be defined as the separation of the sacred from the secular or the church from the state. See McFaul, *The Future of Peace and Justice in the Global Village*, chapter 11, "The Sacred and the Secular and the World Religions;" and Stark, *The Triumph of Christianity*, chapter 21, "Secularization."

14. "What People Do and Do Not Believe In."

"Please indicate for each one [of the categories in this table] if you believe in it, or not." A summary of the results is given in Table 5.

Table 5. American religious beliefs in 2009. *2009 in U.S.*

Do you believe in . . . ?	% yes
God	82
Miracles	76
Heaven	75
Jesus is God or the Son of God	73
Survival of the soul after death	71
The resurrection of Jesus Christ	70

When compared to other surveys, the Harris poll reveals that the response percentages in each of the above areas remain constant over time. As shown in Table 5, belief in God persists among the vast majority of the population—82 percent. Furthermore, more than three out of four respondents (76 percent) reported that they still believe in miracles. Seventy-five percent indicated that they believe in heaven, and 71 percent that the soul survives after death. Clearly, contrary to the forecasts of Voltaire and others, religion has not yet vanished.

Nor is it likely to do so as the following statistics related to the world's large religions show. Among the more than seven billion people who inhabit the Earth, Table 6 gives the numbers of adherents for each one.[15]

Table 6. World's major religions and number of adherents. *hnt'l*

Religion	Adherents
Christianity	2.1 billion
Islam	1.5 billion
Hinduism	900 million
Confucianism and Daoism	394 million
Buddhism	376 million
Sikhism	23 million
Judaism	14 million
Baha'i	7 million

15. "Major Religions of the world Ranked by Number of Adherents."

The above statistical summary of the huge numbers of adherents to the world religions along with the surveys conducted by Harris and other polling organizations reveal that the devotees of all faiths remain dedicated to the spiritual beliefs they learned during their upbringing. Thus, not only do children develop their religiosity early in life through a combination of genetic predispositions and social learning, they carry their allegiances forward throughout the entirety of their lives.

Near Death Experiences

Thus far, in the above three sections of this chapter we have shown that the genetics of the brain in combination with the environment play a role in the development of spirituality, although there is an absence of precise evidence regarding how much each of these two factors contributes to the process. What we can conclude with a high level of certainty is that religious beliefs do not just emerge among children or twins and then disappear in adulthood. The contrary is true. Spirituality persists in the lives and minds of faithful followers well into old age and up to the point of death. This realization leads us to next level of research. Is there any evidence that links the brain to near death experiences? As shown below in the following Tables, the answer to this question is yes.

From K. Nelson[16] we have a summary of a typical sequence of near death experiences where a traumatic event occurs to a person such that person nearly dies but does not. Enough of the person's memory is retained so that a description of the person's reaction to the event can be documented. Such a sequence is summarized in Table 7.

Upon further investigation of people who had near death experiences, researchers found that they went through some common consequences as listed in Table 8.

---

16. Nelson, *The Spiritual Doorway in the Brain*, 101.

Table 7. This is roughly the sequence of what often happens during a near-death experience.

Beginning	Recognizing the crisis
	Feelings of peace
	A noise (buzzing)
	Dark tunnel
	Light
To	Out of the body
	Meeting others
	Being of light
	Life review
	Reaching a border
End	Returning

Table 8. The nature of near-death experience in 55 research subjects, Greyson near-death experience elements.

Category	Consequence	Percent of subjects
Cognitive	Time sped up	62
	Rapid thoughts	44
	Life review	36
	Profound understanding	60
Affective	Felt peace	87
	Felt joy	64
	Felt harmony or unity with the world	67
	Saw or felt brilliant light	78
Paranormal	Vivid sensations	76
	ESP-like awareness	31
	Saw future scenes	29
	Separated from body	80
Transcendental	Entered another world	75
	Encountered mystical being or presence	55
	Encountered deceased or religious spirits	47
	Reached a border or point of no return	67

What provoked these near death experiences? Examples include blackouts (or fainting), heart disturbances, near drowning, and others.

Studies have been made that compare near death experiences with much less serious experiences, such as fainting. See Table 9. Three experiences were much different, comparing near death experiences with fainting: 1) Audible noise or voices was 3.5 times more likely to occur in fainting than in near death experiences. 2) A life review occurred in 32 percent of the near death experiences and not at all in the fainting experiences. 3) Knowledge of the future occurred in 6 percent of the near death experiences and not at all in the fainting experiences.

Table 9. This is based on the study of Dr. T. Lempert and his colleagues, who compared his results with fainting with the near-death experience narratives compiled by Moody.

Experience	Near-death, %	Fainting, %
Out-of-body	26	16
Visual perceptions	23	40
Audible noise or voices	17	60
Feeling of peace and painlessness	32	35
Appearance of light	14	17
Life review	32	0
Entering another world	32	47
Encountering preternatural beings	23	20
Tunnel experience	9	8
Knowledge of the future	6	0

What can we conclude about near death experiences? A careful review of the above Tables leads to the following observations. In Table 5, many of the comments related to near death experiences appear to be associated with the newer part of the brain, such as profound understanding, feeling of peace and joy, the feeling of harmony with the world, and encountered a mystical presence. For the followers of the world's many great religions, these and other similar experiences are associated with the sense of being in the presence of a higher or supreme power of some kind. The older reptilian part of the brain is identified with a distinctly different set of traits, such as anger, fighting, and general aggression.

While no evidence exists that near death experiences provide a glimpse of what might happen after actual death occurs, what is clear is that the brain and spirituality are linked together in the reports of those who have been near death and returned to life. This raises another significant question. Is there a purely genetic explanation for near death experiences that we humans associate with a divine encounter of some kind? In other words, does a God Gene exist somewhere in the neurology of the brain that can explain spiritual experiences as purely physiological events?

The answer to this question operates at two levels. First, there is widespread acceptance among both theists and atheists or naturalists that the experience of human spirituality is dependent on and extends from the physiology of the brain. Without a brain there can be no spiritual experiences. Does this mean that the spiritual experiences that depend on the brain can be reduced merely to the biochemistry of the brain?

This question is similar to asking whether the mental activity we call thinking can be explained solely in terms of the brain's biochemistry. The answer to both of these questions is that nobody really knows for sure. This topic has been discussed extensively elsewhere[17] and will not be repeated here. These two areas remain mysteries of human experience that have yet to be resolved not only between theists and atheists but among atheists themselves. Theists hold that it is entirely possible that divine encounters involve spiritual forces that cannot be explained away by naturalists' assertions that lack scientific evidence. This line of reasoning applies to the thinking process as well, even among naturalists who contend that thoughts cannot be reduced to mere physiological events. Whether and how this mystery will be resolved at some point in the future remains to be seen.

This takes us to the second level: the search for a God gene. This is the topic of the next section.

God Gene

In addition to the research that has been conducted on near death experiences as described above, Dean Hamer, a well-known geneticist at the National Institutes of Health, carried out a special research study to determine how the brain's genetic foundation might be related to or even

---

17. For an extensive discussion of the unresolved brain-mind problem as viewed from the perspective of both science and philosophy, see McFaul, *The Future of Truth and Freedom in the Global Village,* chapter 5, "Roaming around in the House."

explain spiritual experiences.[18] From a purely naturalistic perspective, he asked the following question: "Are our feelings of spirituality caused only by our brain's genetically triggered chemical environment and nothing else?" For purposes of the study, he developed a transcendence test in which he defined spirituality as a process of transformation caused by inner experiences associated with the supernatural. The test involved three components associated with an individual's spirituality. They are 1) loss of sense of self through total absorption in an activity such as reading, 2) a feeling of linkage to a larger universe, and 3) openness to beliefs that are not empirically tangible or provable, such as extra sensory perception.

Hamer assumed that these three components formed a reasonable basis for measuring a person's spirituality. He enlisted about one thousand people to participate in the research project. His testing procedure involved using a questionnaire on transcendence that each participant completed independently of the others. Once finished, Hamer correlated each subject's questionnaire responses with his or her DNA for the purpose of identifying a possible genetic foundation for different spiritual experiences. Hamer reasoned that a high spirituality score might be connected to the presence of a certain gene while a low score might be tied to its absence.

After reviewing these data, Hamer initially believed that he had found such a gene. It was a version of the gene VMAT2 that has been shown to regulate dopamine and serotonin. Dopamine is the brain's "feel-good" chemical that helps runners feel euphoric after a workout. It is also related to amphetamine addiction. Serotonin is associated with modifying psychological moods. For example, it is released by the drug Ecstasy and causes a user to feel high. The drug Prozac, which works more slowly, also alters moods through the release of serotonin.

Despite Hamer's belief that he had discovered the God gene, when comparing the differences between the research subjects' transcendence scores and how they correlated with their VMAT2 genes, he could account for only one percent of the variance. As a result, he concluded, "Clearly this is not the gene that makes people spiritual." Furthermore, he added that "There probably is no single gene. It's one of many different genes and factors that are involved. As a biochemist, I don't expect to solve the mystery of spirituality with one genetic analysis. But I do hope that it will inspire other scientists to get involved in this area."[19]

18. Hamer, *The God Gene*.
19. Quoted in Hagerty, *Fingerprints of God*.

Thus, based on Hamer's pioneering research and his own conclusion, we cannot determine how the brain is connected genetically to human spirituality, to belief in God, or to how this connection causes different people to develop and change their religious views during their lifetimes. For example, another well-known geneticist Francis Collins was an atheist until he turned twenty-eight years of age, when he converted to Christianity.[20] Like Collins, there are thousands of people who change their religious views in many different directions during the course of their lifetimes—from atheism to theism, theism to atheism, and from one form of theism to another. How and why does this happen? The answer is that we do not know. Thus, based on Hamer's conclusion that the VMAT2 gene does not explain the mystery of spirituality, we will not speculate one way or the other about the genetic causes of human spirituality or belief in God. However, what is certain is that genes play a major role along with environmental influences.

Patternicity

The cumulative evidence based on the above studies is that the brain appears to be structured for spirituality. However, one important issue remains to be discussed before we conclude this chapter. It involves the question of whether or not the divine concepts or feelings of being in the presence of a higher power, which are connected with spiritual experiences that occur in the brain, have any objective reality outside the brain. Is there any research that suggests that there is no objective reality to the subjective sensations or perception of the brain? In surveying the current scientific studies related to this possibility, we find that one of the leading candidates was first described by Michael Shermer and can be termed patternicity.[21]

Patternicity refers to the brain's ability to develop patterns that originate from humanity's evolutionary adaptation but have no objective reality, especially as this pertains to religious visions and concepts. There are many examples of people who see images of the Virgin Mary in the sky and similar visions and interpretations. In some cases, how one responds to such experiences could be a matter of life and death. Let us assume that one of our early ancestors heard a rustle in the grass but does not know what causes it. How might he or she react to it?

20. Collins, *The Language of God.*
21. Shermer, *The Believing Brain,* chapter 4 "Patternicity."

Two possible errors could be made in interpreting this sound: Either the rustle is 1) a dangerous predator or 2) just the wind. If our ancestors mistook the sound for a dangerous predator but it was only the wind they became more wary, avoided the area, and lived. This is called a Type I error, a false positive, where the brain believes something is real when it is not. Conversely if one mistakes the sound for just the wind but it is a dangerous predator, he or she probably dies from the attack. This is called a Type II error, a false negative; the brain believes something is not real when it is. Type I and Type II errors are summarized in Table 10.

Table 10. Summary of Type I and Type II errors in the experiences of early humans.

Interpretation	Actual situation	Outcome	Error Type	Error name
Dangerous predator	Just the wind	Avoidance & life	I	false positive
Just the wind	Dangerous predator	Probable death	II	false negative

If one is going to error in this situation, clearly Type I errors that result in survival are preferred compared to Type II errors that probably result in death. Stated differently, false positive errors strongly selected early human and their brains for survival compared to false negative errors.

For a Type I error our ancient ancestors avoided the area and safely arrived at their destinations. After hundreds of experiences like this the survivors passed along this lesson to their progeny who possibly developed a genetic predisposition not to make false negative errors. After approximately hundreds of thousands of generations, each with countless experiences of this type, we humans evolved to the point of interpreting events in our lives as encountering a dangerous predator when it was only just the wind that signified false positives. Our human tendency, then, led us to believe that something was real when it was not. By extension might it be possible that this false belief tendency, which began among our ancient ancestors, evolved into the human inclination to believe in God when in reality God does not exist?

When the concept of patternicity is coupled with humanity's evolutionary adaptation over thousands of generations, it is but a small step to see how someone might see Unidentified Flying Objects (UFOs) in the sky and possibly see divine influences when there might not be any. The accumulation of all those millions of survival experiences might have biased humanity in this direction. We find patternicity is a strong argument for explaining why we humans believe in God when there might not be a God. Thus, while it is clear that the brain and spiritual experiences are closely connected, the assumption of patternicity leads to the conclusion that even though God might not exist, belief in God has contributed to human survival over the long trek of evolution.

Conclusion

We draw this chapter to a close with a brief conclusion of the results that stem from research on the brain and its connection to human spirituality. Based on the above areas of study, at this juncture in the search for scientific knowledge, we lack a foundation of precise evidence of how genetics (nature), the brain (nature), and the environment (nurture) interconnect in the development and persistence of human spirituality from childhood to death. While the combined research findings on 1) how children create their concepts of God, 2) fraternal and genetic twins, 3) the persistence of religious beliefs among adults, and 4) on feelings and beliefs related to near death experiences are strongly suggestive, they are far from complete.

In addition, when we introduce research on patternicity, which holds that belief in God is the result of evolutionary adaptation even though God might not really exist, we are left with one impression: more research is needed before we can reach the level of hard and fast scientific evidence. Nonetheless, despite the lack of definitive knowledge stemming from the above studies of diverse groups of people, as well as the persistence of our inability to explain how the brain and human religiosity are connected through some combination of nature and nurture, existing evidence leads to one overarching conclusion: the human brain is structured for spirituality.

6

Humanity Is Structured for Justice

Introduction

In the previous chapters, we focused on issues related to the central role that knowledge plays in human evolution along with the best available scientific evidence related to the universe and the Anthropic Principle, the emergence of life from non-life, and the brain and spirituality. In this chapter and the next, we shift our attention toward broader social and ethical concerns and away from those that apply more directly to the physical and biological elements of our human nature. In this chapter we will concentrate on humanity's innate quest for justice and in the next of whether or not there exists a universal morality that cuts across all societies irrespective of their cultural or religious differences. Wherever it is appropriate and relevant, we will integrate ideas from the previous chapters into this one and the next.

In the process of analyzing the concept of justice, we will also examine several closely connected concerns. These include the emergence of competition and cooperation during the long trek of human evolution and how these two aspects of human behavior relate to the older reptilian and newer cerebral cortex parts of the brain as we described them in the previous chapter. Once we have completed our exploration of the human search for justice, we will be ready for the next chapter in which we will endeavor to answer the question of whether or not there exists a universal morality, and if so how we might understand it.

The Quest for Justice

By justice we mean fairness or—in the name of fairness—giving individuals what they are due. From the time of Aristotle,[1] the fundamental principle of justice has been interpreted as treating equals equally and unequals unequally. While this distinction might seem confusing at first, a deeper look reveals it is not—as the following examples illustrate. All winners in Olympic running events receive a gold medal. Since all are equal in victory, they are treated equally with identical outcomes.

Second place finishers receive a silver medal. Because their performance is unequal with that of the winners, they are treated unequally. Gold goes to first place and silver to second. In business settings, differences in pay are usually related to dissimilar levels of performance, responsibility, and corporate leadership. The saying "do the time that fits the crime" summarizes the range of unequal sentencing levels that are coupled with felony activities varying from first degree homicide (which receives the harshest penalties) to robbery (where they are less severe).

While these examples demonstrate how dissimilar outcomes lead to different rewards and punishments, applying the fundamental principle of justice across the full range of human activities is one of the most complex challenges that every group faces regardless of its size. Starting with Aristotle's approach to the concern for fairness, the idea of justice has been expanded to include three main types. The first is compensatory justice that involves restoring to a person what was lost when harmed by another; the second, retributive justice, refers to punishing a person for doing wrong as in the above example of criminal behavior; and the third, distributive justice, entails the fair allocation of a society's burdens and benefits to everyone.[2]

Distributive Justice

For the purposes of this chapter, we will focus only on distributive justice because it applies universally to all individuals at all times and in all places and not only to aggrieved persons or groups who seek compensation or retribution after a harm of some kind has occurred. While all three are essential for maintaining a society's well-being, distributive justice takes

---

1. Aristotle, "Nicomachean Ethics."

2. Solomon, *Introducing Philosophy*, 604.

priority over the other two because it embodies the acceptable rules of fair play that determine the allocation of burdens and benefits to all members prior to any wrong doing. Compensatory and retributive forms of justice are secondary because they apply to specific individuals or groups only after an unfair action that violates the rules has taken place.

Despite the priority of distributive justice over the other two along with the easily understood notion that different rewards and punishments should be connected to dissimilar outcomes, determining the precise amounts that each individual or group should get is no simple task for the following reason. Cross-societal comparisons reveal that a pluralism of positions exists. This is because diverse societies use different criteria in deciding how to determine the just allocation of their burdens and benefits. Another way to say this is that the rules of fair play as well as proportioning rewards based on outcomes vary from society to society.

The Four Allocation Principles of Distributive Justice

At the same time, the variations are not infinite—for one main reason. There are only four allocation principles that all societies use to decide who gets what and why. These principles are 1) equal shares, 2) effort, 3) achievement, and 4) need. It is in how dissimilar societies arrange these four in relationship to each other that they differ. Before describing how this works out in the real world, and in order to avoid later confusion, we start by defining these four principles.[3]

By equal shares we mean that a society grants to all its citizens an equal share of specific social goods. One of the best known examples can be found in the US Declaration of Independence, which asserts that all are entitled to life, liberty, and the pursuit of happiness. The US Constitution and its many amendments detail other equal shares as well, such as right to practice the religion of one's choice providing this does not cause harm to others, the right of free speech, right to vote, and so on. At the most basic level, the concept of equal shares identifies specific rights that are inclusive and inalienable. They cannot be taken away.

Within the fundamental principle of distributive justice, the equal shares criterion is the most comprehensive of the four allocation principles because it applies to everyone in equal amounts. The other three do not.

3. Walzer, *Spheres of Justice.*

Rather they are used selectively to determine who is eligible to receive what goods. The second principle is based on effort, which refers to the amount of exertion, energy, or time that one invests in any given task. In an ideal world, greater efforts should lead to larger rewards. However, this is not always the case because of the third principle—achievement, which involves an actual accomplishment or contribution and not merely the effort one puts into achieving it.

There are numerous examples of how the tension between effort and achievement works out in real life. The fastest runners in any given race are not necessarily those who put in the most effort to prepare for it. In the classroom, the hardest working students do not always receive the highest grades. While effort is no doubt related to achievement, which is probably not possible without some minimal amount of it, the most talented athletes and smartest students are the ones who usually outperform everyone else. At the same time, effort pays off at many other levels of achievement even though it does not lead inevitably to the highest outcome in any given field of endeavor. When all the players on a last place little league baseball team receive trophies at their annual banquet, it is not because they won first place but rather because they all contributed to the team effort.

The fourth and last allocation principle is need, which differs substantially from the other three. Need refers to the existence of a physical or mental condition that prohibits a person who is affected by it from access to the opportunities and benefits that any given society normally provides its other members. Wherever the conditions over which persons have no control are present, societies may alter their laws and physical structures in order to provide them with equal access. As a result, it becomes possible for physically and mentally challenged individuals to achieve at the highest level possible according to their abilities and efforts.

This relationship is illustrated in Figure 10, which shows an inverse relationship between physical-mental needs and access to society's opportunities and benefits. As physical and mental disabilities go from lesser to greater, a person's access to opportunities moves in the opposite direction from high to low. When persons have lesser physical/mental needs, they have higher access to society's opportunities and benefits. Conversely, when persons have greater needs, they have lesser access to the same opportunities and benefits. When a society seeks to establish justice based on need, it will respond by providing special benefits for persons who fall in the greater needs category in order to increase their chances of having the same access to opportunities and benefits that persons with lesser needs have.

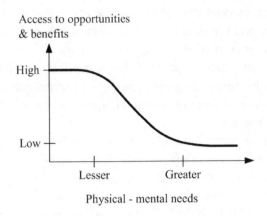

Figure 10. Relationship of needs to access.

Examples are numerous, such as wheelchair confinement related to paraplegia. Unlike the past, in more recent years it is now commonplace for architects to design new buildings or renovate older ones with ramps so that physically disabled persons may enter them. Virtually all societies have created special learning environments for individuals who are mentally challenged with cognitive disabilities of one kind or another—again at an additional cost. The more acute the impairment, the greater is the need for special treatment, as with different degrees of severity involving Down syndrome. Motor vehicles are routinely modified now to accommodate individuals with many different kinds of physical limitations. In 1990, the US government passed the Americans with Disabilities Act that addresses special needs cases of all kinds.

These examples reveal how the four allocation principles of equal shares, effort, achievement, and need relate to both equality and inequality in the distribution of a society's burdens and benefits. As shown, individuals with special physical or cognitive needs are not equal to those who do not have them. Consequently, they should be treated unequally in comparison by being provided with support that levels the playing field of opportunity for all to function normally in society's many social roles. In this way, distributing benefits on the basis of need enhances the principle that all are entitled to an equal share of participation to the extent that this is possible.

At the same time, those who come in first should receive the highest rewards based on their achievements. While the quality or quantity might differ from field to field, the principle of allocating the highest reward applies equally to all top performers. In a humorous but discerning nod to the principle of effort, it has been said that eighty percent of success in most jobs involves just showing up on time. Only twenty percent entails doing the actual work. While this tongue-in-cheek image is no doubt exaggerated to make a point, nonetheless, along with the other illustrations, it captures well how all societies struggle to resolve the challenge of distributing burdens and benefits, that is, justice, according to a widely shared acceptance of how to combine equal shares, effort, achievement, and need. Furthermore, over time, each society will weigh differently the importance of each of the four in relationship to the other three.

For example, some societies like the United States place top priority on market place achievement and to a lesser extent on need. To foster this primary emphasis, every state in the US provides all children between ages six and sixteen with a mandatory equal share of opportunity to receive an education through the twelfth grade—either public or private. In turn, this stimulates many individuals to achieve through their own efforts at even higher levels of education in order to obtain access to rewarding careers and high monetary gains.

At the same time, while states distribute some shares equally such as education for all children, they do not do so in other areas, especially when they are controversial. While all states permit legal marriage between a man and a woman, a growing number have extended this right to same-sex couples. Just how far the US will go in applying an equal share of monogamous marriage to everyone regardless of the sexual preferences of the marrying partners has not yet been determined.

Other societies, such as the European nations, seek greater balance between effort, achievement, and need by offering more generous cradle to grave safety net programs than those found in the US. In many developing regions of the world, as in parts of Africa and Latin America, the uppermost demand for justice involves the need to reduce poverty without which the other allocation principles of equal shares, effort, and achievement find little room for expression.

As all societies over time strive to achieve the highest level of justice that is possible given their unique circumstances, the tension that exists among the four allocation principles offsets the drift toward extremes. For example, a society that overemphasizes the need principle runs the risk of undermining effort and achievement without which no society can progress. A society that places excessive emphasis on individual achievement will ignore challenges faced by those with special needs. If the members of a society perceive that too much weight is being given to one of the principles at the expense of the others, they will take actions necessary to achieve greater balance among all four.

Given the various ways in which diverse cultures combine shares, effort, achievement, and need as they strive to find a place for all of them, the question arises as to whether or not it is possible to develop a scale of justice to determine how much success or failure different societies have had in achieving it and how this might be changing over time. Can we determine that some societies are more just than others? Before detailing an answer to this question, which we believe is yes, it is necessary to explore more deeply how the four allocation principles grow out of the quest for justice. We believe that one of the best ways to grasp this relationship is to examine the role that competition and cooperation play in different societies as they aspire to achieve maximum justice.

Competition and Cooperation

For the sake of clarity some definitions are necessary. By competition we mean comparing alternative positions or persons against each other in order to determine which or who is superior. Contention over some perceived good is the essence of competition. The outcome of any rivalry is summarized aptly in the well-known adage: To the victor go the spoils. Victory can involve a single winner and loser or multiple winners who rank along a graded scale from first, second, third, and so on, as well as losers who fall below the lowest rank. Competition can come in many forms that range along a continuum of friendly at the one end, as in the case of games and sports to savage at the other, as in war and other forms of violent behavior. Friendly competition does not involve taking the life of opponents through killing, whereas savage competition does.

We define cooperation as the opposite of competition. It refers to collaboration between two or more parties. It connotes an image of teamwork and mutual support among members of a single group or between groups. Cooperation occurs when two or more persons perceive they must work together to achieve a common good. The ideal outcome of cooperation is captured aptly in the saying: We're all in this lifeboat together. Let's all row in the same direction. The members of any given cooperating group can range in their views along a continuum from complete agreement to substantial disagreement over any given means or ends that cooperation is intended to produce. Wherever they fall along this spectrum, they are willing to compromise in order for the group as a whole to succeed in any given undertaking that requires cooperation.

In addition to defining the difference between competition and cooperation, other terms need to be clarified in order to avoid later confusion. The first is self-interest. By self-interest we mean the motivation that lies behind every person's and group's desires to achieve their goals—however defined. The second term is selfishness, which we define as expressing self-interest narrowly in order to achieve one's own objectives without regard to the consequences for others. The third is generosity, which is often called altruism. It stands for expressing self-interest broadly so that it includes not only the achievement of one's own goals but those of others as well. Selfishness is exclusive, and generosity is inclusive. See Figure 11.

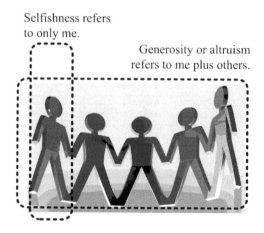

Figure 11. Self-interest can be "selfish" or "generous."

We believe that all human actions emanate from self-interest and that it is impossible for any person or group to act apart from it. It is a contradiction to say that persons and groups can be devoid of interest in the goals they are pursuing. The only real issue is whether self-interest takes the form of selfishness or generosity or some combination of the two. This implies that all expressions of self-interest range along a scale with selfishness at the one end and generosity at the other. It is because self-interest and selfishness are often used interchangeably and contrasted with generosity that we find it necessary to keep the differences between self-interest, selfishness, and generosity in mind. In addition, all three have a direct connection to the other two terms defined earlier—competition and cooperation. How so?

Both competition and cooperation involve the expression of self-interested individuals or groups who can focus narrowly on their own selfish goals or broadly on goals that generously include others as well as themselves. While the combinations involving competition, cooperation, selfishness, and generosity in the expression of self-interest are numerous, the following well known examples highlight how they are entwined.

First, in all cultures, when parents encourage their children to share toys, they are stressing the importance of cooperation among siblings and playmates. Adults show their self-interest when they help children discover and develop their own innate aptitudes and talents that they will express in a self-interested manner as they mature. Ongoing community life would not be possible apart from the contributions that each self-interested individual brings to it. In addition, all persons from children to adults can choose to express their self-interests alternatively through narrow selfishness or more broadly through generosity that includes the self-interests of others.

Second, all group sports from neighborhood based, youth baseball to professional World Cup Soccer involve competition between teams to see who is superior. All players must cooperate with each other in order for the team to be competitive and win. Players must also compete against each other and demonstrate who has the top skills necessary to hold any given position. Each team expresses its narrow self-interest in aspiring to win while being competitive in relationship to other teams and cooperative within itself. Despite the competition, at a broader level, all teams cooperate by following the rules of the game that are designed to promote the good of the sport as a whole within which teams compete to win.

The third and last example involves an area of human behavior that combines competition and cooperation as starkly as any other—and with horrifying consequences—war. Whereas raising children and participation in sports represents the more benign end of the competition—cooperation continuum, war epitomizes the savage side of human behavior. While cooperation and compliance with the chain of command within any given military unit does not guarantee victory over enemies, lack of cooperation among its members almost always guarantees defeat. Internal cohesiveness by following the chain of command is an essential aspect of one of humanity's oldest and most destructive forms of intergroup competition that spins downward in the direction of violent hatreds and hostilities.

In addition to the paradoxical nature of cooperation and competition under conditions of warfare, the outcomes of military conflict can range from narrow to broad self-interests of warring parties. For example, during the American Civil War of the 1860s, the Confederacy's goal of withdrawing from the Union in order to maintain its system of slavery was an expression of narrow self-interest. On the other hand, the Northern army's defeat of the Southern forces embodied broader self-interests because it kept the entire nation intact.

President Lincoln's Emancipation Proclamation that freed the slaves by abolishing slavery, along with the thirteenth through the fifteenth amendments to the US Constitution, opened the door of equality for both black and white citizens. In another example, the broad-based Allied assault against Nazism in the 1940s resulted in the end of Hitler's narrow self-interest vision of Aryan domination and extermination of the Jewish population.

What stands out in both of these extreme examples is how the combination of competition, cooperation, selfishness, and generosity played out as the fight to the death factions battled each other over whose self-interest would prevail. Furthermore, while not as immediately apparent as who won and who lost, the quest for justice played a central role in the North's victory over the South in the US Civil War and the Allies' defeat of Nazism in World War II. Extending equal rights to all is consistent with the allocation principles of equal shares, effort, and achievement.

Competition and Cooperation in Human Evolution

Given the above examples of the many ways in which competition and cooperation are coupled together, how did they become interwoven during the course of human evolution on Earth? The answer to this question is no doubt very complex, but we believe that it can be found by examining the key factors that enabled human communities to survive during the long trek of evolution. We start this section by referring back to the description of the brain's development as presented in chapter 5 and then show how this relates to modern views of human evolution.

The most recent and credible scientific evidence indicates that the Earth is about 4.6 billion years old and that the earliest single-cell life forms appeared around 3.85 billion years ago as discussed in chapter 4. As life on Earth evolved, many species arose, prospered for varying periods of time, and then disappeared—the most spectacular being the extinction of the dinosaurs sixty-five million years ago. As the Earth's evolution continued, the best evidence points to the emergence of anatomically modern humans around 200,000 years ago. Since then the human species in all its combinations of cultures and colors has spread steadily around the world.

Starting in 1859, Charles Darwin and colleagues developed the concept summarized in the phrase survival of the fittest, which addresses the issue of why some groups died out and others survived during the long-haul of the Earth's evolution.[4] After a careful study of the different types of plants and animals that exist on the Galapagos Islands and nowhere else on Earth, Darwin concluded that each species' survival potential depended on its ability to adjust to its surrounding environment. In this sense, what holds true for birds in the air and fish in the water also applies to humans on land. Through distinctive but different patterns of adaptation, each surviving species finds a unique niche on the evolutionary chain. If they fail, over time they cease to exist.

This implies that humanity would not have survived without carving out a special place for itself amidst the Earth's countless other life forms. The one factor, above all others, that made this possible was the development of the newer part of the human brain, that is, the cerebral cortex, because it gave humanity the capacity for abstract thought, religion, language, speech, and memory. In chapter 5, we discussed the evolution of the human brain from its older reptilian origins to newer and higher capacities for abstract thoughts that are centered in the cerebral cortex. In turn, this led to the creation of a shared culture and group solidarity. See Figure 12 in which we provide a summary of our suggested relationships between the reptilian/older and newer part of the brain and their relationships to ethics and justice.

4. Darwin, *Origin of Species*; and *The Descent of Man*.

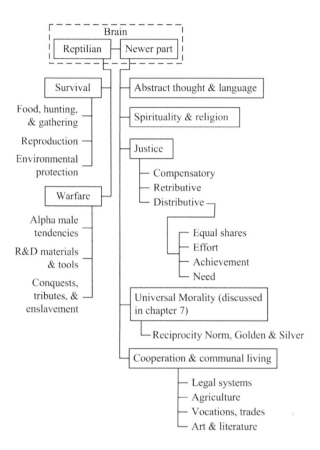

Figure 12. The suggested relationships of the reptilian and newer parts of the brain.

This does not mean that group cohesion is unique to humans. The contrary is true. It applies to virtually all other species. What makes humanity different from other life forms is that humans connect through language and culture, and not merely biological instincts or impulses. While humans share much in common with the other species, such as satisfying the biological need for food, water, and reproduction, and so on, and other species communicate through sound and movement, only humans interact through culture at the highest levels of linguistic abstraction and organization. It would not be an overstatement to say that because we humans created culture we learned not only how to find a niche and survive but to thrive as well—all of which became possible as the newer part of the brain, the cerebral cortex, developed.

As *Homo sapiens* advanced through many millennia of trial and error experiences, thousands of groups splintered off from each other and fanned out across the Earth's dispersed geographical regions. As a result of high mountains, wide oceans, and nonexistent contact, communities became isolated from each other. Then, due to centuries of physical isolation, they developed dissimilar languages and cultural practices. Now, as the world advances into the globalizing stage of human evolution driven by worldwide electronic communication and mass transportation, as described in chapter 2, an increasing number of people across all societies are becoming aware of the degree to which 200,000 years of human evolution have produced such a broad array of diverse cultures.

As humans evolved and became more sophisticated at communicating through language and in devising strategies that insured collective continuity, two factors in particular emerged as essential to their survival. They are the selfish impulse for competition (summarized by: What is in it for me without regard to others?) that is centered in the older area of the brain and generous cooperation, located in the newer part. While these appear to be opposed to each other, they are not. Instead, their relationship is paradoxical and complex. Much of the confusion over the role that each factor plays in the evolution of life on Earth stems from an initial and persistently lopsided interpretation of what Darwin meant by the phrase survival of the fittest.

When Darwin first used this saying to summarize why some species survive and others perish, many interpreters misconstrued his meaning. By survival Darwin meant that successful species find ways to adjust to their natural environment by adapting and finding special places for themselves. While competition among species no doubt plays a role, Darwin believed it was subordinate to the primary challenges associated with successful adaptation. Fish do not compete with birds, and silver back gorillas do not fight giraffes to survive. They all persist separately because each species acclimated to the physical and environmental difficulties it confronted. Each found a unique niche on the evolutionary ladder.

In addition to progressive adaptation, long term survival rests on intra-group cooperation as well as inter-group competition. Bees survive because individual bees know instinctively their role within the hive. Birds fly in specific formations as determined by their cooperative evolutionary adaptations. Dolphins swim together as members of oceangoing schools. While the behavior of most nonhuman species is driven by innate instincts, in humans it is caused by rational choices that occur within the cerebral cortex (nature) and in the context of culture (nurture).

Thus, even though humans compete among themselves, and all too often violently as war and terrorism demonstrate, long term species survival depends on cooperation as much if not more than competition. Thus, based on scientific discoveries related to brain research and culture studies we can conclude that it is a combination of both the older and newer parts of the brain that accounts for this interaction with the cooperation tendencies that stem from the newer part of the brain gaining the edge over competition in fostering human survival and adaptation.

In light of this evidence, it is now clear how competition, cooperation, and distributive justice became joined together in human evolution. Sharing burdens and benefits among all members of society (distributive justice) enhances the norm of cooperation in doing what is good as well as avoiding what is harmful. This does not mean that there is no role for competition in the application of distributive justice. The contrary is true. As the earlier example of activities ranging from youth sports to professional soccer makes clear, competition stimulates the pursuit of excellence in a broad range of activities, including those related to economic and political competition as well.

At the same time, when competition leads to violence and war, it crosses the line from enhancement to the destruction of life. While both competition and cooperation may lead to an equitable sharing of society's burdens and benefits, cooperation holds the edge because it increases generosity and concern for others in the expression of self-interest. In turn, this ties to the four allocation principles of justice. The desire to cooperate leads a society to expand an equal share of rights to all its members, stimulates persons to exert more effort to achieve common as well as individual goals, and motivates them to reach out to help those with special needs.

A Comprehensive Theory of Justice

Given the diverse ways in which societies emphasize some of the allocation principles more than others in the quest for justice, is it possible to develop a theory of justice that combines them all in a systematic way? The answer is yes, and the one name that stands above the rest in moving in this direction is John Rawls who developed a systematic modern contractarian view of distributive justice. The contribution that he has made over and above everyone else in developing a systematic theory of justice in the past fifty years cannot be overstated.[5] Whereas during most of history issues related to fairness in the distribution of life's burdens and benefits were addressed in terms of theologically oriented or natural law theories, by building on the ideas of earlier contractarian writings that started in the mid-seventeenth century with the works of Thomas Hobbes, it was Rawls' approach that laid the foundation for the current social contract perspective.

Like other writers before him, Rawls starts by defining justice as fairness. From this point of departure, he then adds a crucial component that sets the stage for how justice applies directly to social cohesion. He assumes that any society's internal harmony depends on the extent to which its citizens believe that burdens and benefits are being distributed equitably. Perceptions of fairness strengthen social unity, and unfairness weakens it. For Rawls, distributive justice and social stability are inseparable.

This implies that the internal stability of every community, from small groups to society as a whole, depends on the degree to which its members perceive that the distribution of burdens and benefits is fair. The potential for instability increases when there is a widespread perception that advantages go disproportionately to some while afflictions fall hardest on the shoulders of others. This is especially so in the case of widespread discrimination where individuals and groups are prohibited from getting their fair share of the benefits because of demographic factors such as race, gender, sexual preference, religion, ethnicity, national origin, and so on, which have nothing to do with their innate abilities.

When individuals and groups become convinced that the distributive justice system in which they live is unfair to them, then the probability increases that they will reject the social arrangements that perpetuate it. In response, they will do what is necessary to overcome their deprivations even if this leads to crime or to efforts to change the system through legitimate political channels or open rebellion. In other words, despite all of the complexities associated with the establishment of distributive justice in any given society, social stability is highly correlated with the belief that the burdens and benefits are apportioned fairly across the boundaries that separate individuals and groups from each other.

Figure 13 illustrates this relationship. When levels of justice move from less to more, a society becomes more stable. Conversely, when a society moves increasingly toward less justice, it becomes less stable.

5. Rawls, *A Theory of Justice*.

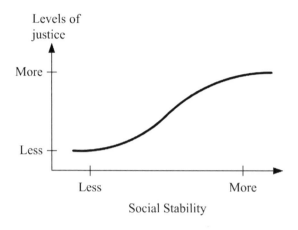

Figure 13. Justice and social stability.

Given the diverse ways in which different societies weigh the four allocation principles of equal shares, effort, achievement, and needs against
each other, Rawls is keenly aware of the challenges that are associated with
developing a comprehensive method for measuring how levels of justice
vary from society to society. At the same time, he does not accept the
ethical relativism that this implies. He rejects the notion that one pattern
of justice is just as good as another. He moves beyond the relativist option
by developing principles that can be used to measure degrees of fairness in
all societies and to show the extent to which they vary. While not everyone
agrees with him, his approach has become the norm in current discussions
dealing with distributive justice. It is not far from the mark to say that on
the subject of distributive justice for the past half century all paths lead to
and from Rawls.

He starts by asking us to imagine in the form of a thought experiment what it would be like in an "original position" where social patterns
or cultural norms have not yet emerged. In addition, everyone would stand
behind a "veil of ignorance" in not knowing whether they would be born
into a rich family, a poor one, or somewhere in between. They would not
know their birth order, sex, color, nationality, religion, their parents' education levels, positions, occupations, and so on. Nor would they know how
their lives would progress, what they would do in life, how long they would
live, and how, when, and under what circumstances they would die. The
goal of this thought experiment is to develop a set of justice concepts with

which everyone would agree in order to progress through life under conditions of maximum fairness.

Rawls knows that this is not how life plays out in reality given that everyone is born into dissimilar life circumstances and with different abilities. Nonetheless, he articulates a set of principles that he believes everyone would accept as a way to achieve maximum justice if humanity could start all over again—so to speak—prior to the course of history as it unfolded during the past several millennia. In the process, he combines the four allocation principles into a systematic method for comparing the levels of justice that exist from society to society.

Building on the ideas of the original position and the veil of ignorance, Rawls goes to the next level by describing the two substantive ideas that are the heart of his theory of justice. They are 1) the liberty principle and 2) the difference principle, which contains two parts. Part 2a deals with the inequalities that emerge inevitably in every society and how to know when they are just or unjust. Part 2b consists of a principle of fair equality of opportunity.

The liberty principle is identical to the equal shares concept that was discussed above as one of the four allocation principles associated with distributive justice. In Rawls' theory of justice, liberty means that each person has an equal right to the basic liberties granted to all persons. No one is excluded. For Rawls, liberty takes priority over all other principles because if equal rights, or shares, do not include everyone, then justice is not possible. For example, injustice occurs when only a limited segment of citizens has the right to vote, to receive an education, to practice the religion of their choice, and so on.

Acceptance of the liberty principle means that everyone starts their chance to succeed in life on the same moral and legal footing—as equals. At the same time, Rawls recognizes that individuals will be born into different families and social classes, have diverse talents and attributes, have dissimilar levels of aspiration and opportunities, and vary in the work habits that lead to unlike outcomes. It is at this point that he makes one of his most creative contributions.

While Rawls is an egalitarian on the subject of common liberties for all, he is not an egalitarian on the issue of end results. He accepts that effort and achievement levels will differ from person to person, which in turn will lead to an unequal distribution of society's burdens and benefits. Given his recognition that inequality of outcomes is as much a part of life as the equality of liberties for all should be at the start of life, the only real questions are 1) whether unequal outcomes are just and 2) how to determine this.

The second part of Rawls' theory—the difference principle—addresses these two questions. Just as his primary emphasis on liberty (equal shares) could easily be designated the equality principle, his concern for assessing the status of different outcomes (due to varying degrees of effort and achievement) could be called the inequality principle. In other words, his theory of justice encompasses both equality and inequality and defines the conditions under which they are either just or unjust.

Unequal outcomes or differences are just if (2a) everyone benefits from them, and if (2b) equality of opportunities continues to remain open for all. For Rawls, the difference principle balances the liberty principle, and both include the four allocation principles in different ways. Expressions of liberty (based on equal shares) should not benefit only the few but the many; and the direction of social change should lead over time to improving the conditions of the less fortunate (based on need). As opportunities open up for the less fortunate as measured by long term statistical comparisons, it can be said of the society that is moving in this direction that it is becoming more just. When liberties wither, poverty increases, and equal opportunities shut down, a society is becoming increasingly unjust.

For example, when monopolies evolve out of competitive business, they dominate their markets to such an extent that they can prohibit others from entering them. This is a condition of injustice because the initial expression of liberty (the right of all to compete in an open market place) leads to the loss of liberty (the monopolistic foreclosure of equal opportunities for all to compete). In the realm of economics, when the number of noncompetitive monopolies increases, so does injustice. On the other hand, when competition among firms continues to grow and allows more people to engage in the production of high quality goods at cheaper prices, which in turn benefits consumers, then justice increases.

A word of caution is in order here. It must not be assumed that Rawls is advocating a particular type of political or economic system because all of them possess the potential to be just or unjust. Rather, he is identifying the justice principles that can be used to compare many different types of societies to each other. Those societies 1) with the most extensive number of equal liberties, 2) with unequal outcomes that benefit everyone—especially those who suffer the greatest hardships, and 3) that continue to retain the greatest number of equal opportunities for all are the most just.

In addition, if Rawls is correct that the degree of any society's internal stability is tied to its citizens' perceptions of the extent to which burdens and benefits are distributed fairly, then those societies with the greatest amount of justice will have a long term survival advantage over those that do not. When a small political class dominates the many, when economic wealth is concentrated in few hands, when poverty is widespread and a middle-class does not exist or begins to shrink in size, and when liberties and equal opportunities are not present or dry up over time, a society can be said to be unjust or is becoming progressively more unjust.

As a society's level of justice decreases, it will become increasingly unsettled. In turn, this will strengthen the odds that it will drift in the direction of social disruption or even open rebellion by the masses, which the state will counter with police and/or military action. Internal turbulence will become wide spread, and the potential for violence will increase on both sides with no guarantee of how this will play out in the short run. In the long run, however, those societies that survive, transition out of internal crises caused by injustice, and increase the total amount of justice in the process will prevail over those that do not.

History is filled with many such examples. One of them has already been discussed previously, namely, the American Civil War of the 1860s that led to the Northern victory over the South and the elimination of slavery. In addition to the massive amount of killing that occurred as Federal and Confederate troops slaughtered each other during this brief time period, the passage of the thirteenth—fifteenth amendments of the US Constitution paved the way for other groups to seek an end to discrimination that for decades had locked them out of privileges that the majority took for granted. Building on these earlier amendments, the nineteenth amendment, which was ratified in 1920, finally extended to women the right to vote 144 years after the nation was founded in 1776.

In the language of justice, the amendments to the US Constitution reveal that equal shares, which Rawls calls liberties in the first principle of his theory of justice, have been expanded to include an increasing number of historically excluded groups such as racial, religious, and ethnic minorities, women, and the disabled. The current thrust by some segments of the US to extend marriage rights to same sex couples is an example of the latest, and probably not the last, initiative by another historically excluded group to stretch the boundaries of liberty in its direction. Only in the future will it be known how far this growing initiative will go. In terms of Rawls' difference principle, the above examples show that over time the US has become a more just society as a long line of historic outsiders have followed each other through an ever widening window of liberties and equal opportunities once denied them.

In addition, global long term trends during the past 200 years have moved in the direction of a social contract view of distributive justice as an increasing number of nations have adopted democratic forms of government based on the idea of popular sovereignty that emerged in the 1700s. In the year 1840, only two democratic nations existed—The United States and England (3.9 percent of the Earth's population). In 2000, 57.1 percent of the world's populace was associated with democratic societies. This is plotted in Figure 14, which shows the remarkable growth in democracies since 1840.[6] Most recently, popular uprisings during the early twenty-first century in countries such as Egypt, Libya, Tunisia, among others demonstrate a desire of citizens in historically autocratic countries to join the long term trend of establishing their own democratically elected governments throughout the Middle East.

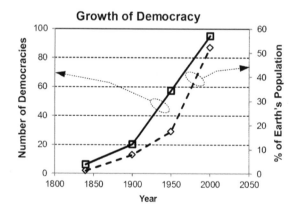

Figure 14. Number of democracies and percent of Earth's population.

Many national economies have also felt the effects of the liberty and equal opportunities principles as Rawls defines them. With the collapse of global communism in 1990 along with its centralized systems of control, the majority of the world's economies, including China, have become

6. McFaul, *The Future of Truth & Freedom in the Global Village*, 150–51.

regulated market places that foster entrepreneurial initiative (based on equal shares, effort, and achievement). At the same time, a majority of nations have created safety net programs that protect marginalized or disadvantaged groups (based on need). While the balance between government support and regulated market place freedom varies from society to society, the time line is moving in the direction of increasing the total amount of political and economic benefits for a growing number of people throughout the emerging global village—as measured by Rawls' liberty and difference principles.[7]

In sum, while the world still has a long way to go to become a distributive justice utopia, nonetheless, we can conclude that while some nations lag behind others and access to liberties and equal opportunities varies across societies, the long term trend reveals that the nations of the world are becoming progressively more just.

Conclusion

This chapter explores in depth the idea of justice, which we defined as fairness in sharing society's burdens and benefits. Following Aristotle's widely accepted views we laid out the fundamental principle of distributive justice, which involves treating equal achievements with equal benefits and unequal outcomes with unequal rewards.

We noted that while this idea is easy to grasp, implementing it across the full range of human activities is anything but simple. This is because different societies distribute burdens and benefits in diverse ways depending on how they define the rules of fair play as well as who should get what and why in any given situation. Achieving the fundamental principle of justice in any society depends on how it distributes burdens and benefits according to the four allocation principles of equal shares that apply to everyone along with differences that exist between individuals and groups according to their dissimilar efforts, achievements, and needs. When compared against each other, societies differ in the ways in which they give these factors dissimilar degrees of emphasis.

Some, like the United States, place a high priority on achievement. Others, such as the European nations, have developed a wide array of

---

7. For an expanded description of the major political and economic changes that have occurred in the modern world in comparison to patterns that existed during the pre-modern era, see McFaul, *The Future of Truth and Freedom in the Global Village*, chapters 8 and 9.

safety net programs designed to promote the welfare of needy populations from birth to death. Because of poverty, many of the developing nations throughout Africa and Latin America focus on the needs of the disadvantaged, which takes precedence over emphasizing the other three allocation principles.

Next, in order to understand more deeply why all societies include all four allocation principles rather than one or fewer than four in their quest for justice, we examined the underlying tension between competition and cooperation that exists in all societies. We observed that when individuals and groups compete against or cooperate with each other, they do so in order to advance their self-interests, which range along a continuum from selfishness at one end to generosity at the other and with various combinations in between. Since no individual or group can act apart from self-interest, the only issue is whether their actions involve narrow and selfish consequences for the few or broad and generous ones for the many.

Furthermore, the mix of competition and cooperation within human as well as non-human conduct is readily explainable in terms of the Darwinian view of evolution. Long term evolutionary survival includes both. Those species that can adapt successfully to the challenges posed by their physical surroundings are the ones that stand the greatest chance of surviving. When species cannot adapt or hold their own against those that outcompete them, they increase the odds of becoming extinct. In the final analysis, the survival potential of any species depends on its ability to find a unique niche on the evolutionary ladder.

As species struggle to survive, they engage in both competition and cooperation as means to this end. In order to outlast competitors all species must devise ways for their members to cooperate together in order to overcome threats to their ongoing existence. On the evolutionary chain, nature has provided the fittest nonhuman species with natural biological instincts that direct their behavior, as in the case of bees and seagulls.

Humans evolved differently by developing a two level neurological system that includes both older and newer parts of the brain. The evidence from brain research indicates that the stimuli for competition and conquest reside in the older reptilian part called the autonomic nervous system. The newer and higher part is called the cerebral cortex, which fosters the development of memory and speech and stimulates the capacity for cooperation.

With the emergence of the cerebral cortex, humans gradually began developing the capacity to create and live in culture based on rationality and cooperative solving abilities in addition to similar inborn instincts with which nonhuman species are endowed by nature at birth. For humans, competition and cooperation are connected to the two distinct parts of the brain, and both provide the basis for humanity's quest for justice. Through effort and achievement, the need to compete and win is satisfied. The allocation principle of equal shares involves the cooperation of all members of society in extending rights to all and to agreeing to meet the needs of disadvantaged individuals and groups that require special support.

Building on the ideas of Aristotle's fundamental principle of justice and the four allocation principles, in the past fifty years John Rawls more than anyone else has contributed to the formulation of a modern day theory of justice. He is credited with developing a systematic social contract framework that is useful for comparing the levels of justice that vary from society to society. He starts by imagining what an agreed upon set of principles might look like if humanity could start history all over again with the goal of striving for maximum fairness for everyone.

In the process, Rawls reduces and combines the four allocation principles of equal shares, effort, achievement, and need into only two—the liberty and difference principles. Liberty parallels the equal shares principle, and the two part difference principle combines all four. Two-a of the difference principle holds that unequal outcomes in distributing burdens and benefits through effort and achievement are just if society as a whole benefits, specially the least advantaged (the needy). Two-b maintains that the equality of opportunities should remain available to all (equal shares) even as different individuals and groups compete (effort) to receive society's highest rewards (achievement).

Those societies that establish, preserve, and expand liberties to formerly excluded populations as well as serve the needs of down trodden groups while maintaining equal opportunities for everyone are the ones that are the most just. There is ample room for both competition and cooperation in the expression of self-interests. As the global village of the twenty-first century and beyond continues to expand, it will be cooperation among diverse groups more than the desire of a few to conquer the many that will increase humanity's potential for long-term survival and progress. As the evidence makes clear, this is so because humanity is structured or tuned to pursue and maximize the attainment of justice for all.

7

Humanity Is Structured for a Universal Morality

Introduction

In this chapter, we will examine the issue of whether or not a universal code of ethics cuts across the diversity of world cultures and how it relates to the concept of justice as described in the preceding chapter. Our search for the existence of such a morality will also include the opposite possibility that each culture's moral standards are relative and self-contained with little or no overlap with those of other societies. We are using the term universal morality to refer to the existence of a core set of values that can be found in all societies irrespective of differences in their cultural practices. Moral relativism implies the opposite, namely, that no such universal core of values exists.

In the discussion that follows, it will become clear that opinions differ on this subject. Anyone who spends even a small amount of time searching for a global morality will encounter others who are convinced that this is a futile effort. In their view, moral relativism, and not moral universalism, is the dominant pattern that separates human beings from each other. Nonetheless, we will show that despite the existence of an enormous range of cultural practices that separate societies from each other, there sits at the center of all cultures a universal morality; and that moral relativism cannot be supported by the empirical evidence.

Ethics and Morality

First—definitions: In the field of Ethics, the terms ethics (or ethic) and morality are often defined differently, and no consensus exists on how to differentiate them from each other. Some authors define morality as standards of right and wrong and ethics as the theoretical justification for them.[1] While this distinction helps in exploring moral behavior in applied fields such as bioethics or business ethics, we will not follow it in this chapter. This implies that in the search for a universal morality, we will use the words ethical and moral interchangeably because both center on the human quest to identify common values that enhance to the greatest extent possible the overall goodness and well-being of all societies. However, in order to avoid confusion, throughout the remainder of this book, we will employ only the term moral universalism when comparing and contrasting the search for a universal morality with the moral relativist position.

Moral Relativism and Cultural Relativism

Additional definitions are required before we proceed farther. Like the terms ethics and morality, the potential for confusion falls on other concepts as well especially over the difference between moral relativism and cultural relativism. While often seen as synonymous, they are not the same for the following reasons. Social scientists use cultural relativism to refer to the science of describing the normative differences that exist between societies without making judgments regarding the superiority of one over the others.[2] (Here the word normative refers to setting a standard or norm.) While they might have personal preferences, as scientists they suspend their opinions when writing about and comparing intercultural moral differences and preferences.

Moral relativism is different. It is an ethical theory and not a scientific method for studying human behavior. When anthropologists or sociologists suspend their moral judgments during the course of describing different cultures, they are employing in a correct and consistent manner the scientific rules associated with cultural relativism.[3] If, however, they

---

1. Munson, *Intervention and Reflection*, 860–915. Velasquez, *Business Ethics*, 3–27.

2. Boaz, *The Mind of Primitive Man.* Kluckholn, *Mirror for Man.* Kroeber, *Anthropology.*

3. Benedict, *Patterns of Culture.* Mead, *Coming of Age in Samoa.* Mead, *Sex and Temperament.*

take the next step and assert that the evidence supports the conclusion that moral universalism does not exist, they have stepped out of ethnographic science and into philosophy. In a word, they have become moral relativists. The reader should be aware of this subtle shift, because it goes beyond science and assumes that ethnographic descriptions of diverse cultural norms lead inevitably to a moral relativist deduction. However, as we will demonstrate later in this chapter, this inference is false.

Moral Relativism and Ethical Subjectivism— Individual and Group

Moral relativism also goes by another name: ethical subjectivism, which can be divided into two main branches. One emphasizes the role of the individual, and the other the group. When all is said and done, according to the adherents of both branches, all that can be known for certain is that individuals and groups differ because of the subjective choices they make in creating their own norms. No universal moral framework exists for any person or community to pass judgment on the ethical practices of others. Only persons who are on the inside—so to speak—of an individual's or group's normative system can critique it and not those who stand on the outside of it.

For ethical subjectivists who focus on the individual, moral choices can be reduced to a person's opinions. The chief criterion for determining whether a moral standard or practice is acceptable, and therefore legitimate, rests only on the subjective feeling of believing it to be so irrespective of the opinions of other individuals or groups. In this sense, everyone's preferences are equal in value and no comparative assessment can be made in establishing the relative worth of one set of moral principles over others.

The second branch of ethical subjectivism shifts the emphasis from the individual to the group. Subjectivists who ascribe to this position start with the assumption that no person lives separately from others or acts in isolation. Rather, all persons function first and foremost as members of a community or—better stated—several communities. This implies that an individual's moral preferences are always an extension of his or her group affiliations.

For example, as Aristotle noted centuries ago, at the most basic level humans are by nature social beings. In a similar vein, an ancient German adage captures well this collective outlook: *Ein mensch ist kein mensch.*

(One man is no man.) Because of the process of socialization, as individuals mature during their lifetimes, they internalize the moral standards of the communities into which they are born and through which they grow into adulthood and old age. In a word, every person's values mirror group relationships. Thus, while ethical subjectivists can be divided according to whether they emphasize the individual or the group in the formation of value preferences, they are joined at the hip—so to speak—by their shared rejection of moral universalism.

Moral Relativism in Practice

How does moral relativism play out in real life? While many examples can be used to answer this question, we suggest that one in particular, which we have already considered in earlier chapters, stands out above the rest: Hitler's Third Reich. Germans who joined this regime during the 1930s and 1940s gave voice to belief in the superiority of Aryan values and the anti-Jewish hatred it spawned. It was Hitler's obsession with this worldview and ethical system that fueled his fixation to rule the world, which in turn propelled him and his Nazi followers to invade neighboring states for the purpose of conquest and domination and to launch a campaign of Jewish genocide.

The history of how the Allied Nations rallied, collaborated, and destroyed the Third Reich and its vision of violent global conquest is well known. Books that tell this story typically emphasize the strategies the Allies developed to triumph over Hitler's war machine. At the same time, this narrative is more than just a description of political alliances and military tactics. At its core, it is about the struggle over values, which culminated in the Allies' defeat of an autocratic, fascist morality that the German Nazis sought to impose on the world through hatred and killing.

While the Third Reich represents an extreme case of human cruelty, it would not be an overstatement to say that most of humanity's major hostilities turn on disagreements over ethical concerns and consequent power struggles to determine which persons, positions, or groups should prevail. While not all value conflicts erupt into violence or lead to warfare on par with the massive devastation of World War II, nonetheless, the outcome of moral disputes often cause major shifts in social behavior or produce conflicts that cost thousands of lives.

Another powerful example that we have used above also illustrates our point. The struggle to abolish worldwide slavery along with the moral justification for legally owning, selling, and subjugating human beings to a lifetime of coerced servitude was long and tumultuous. It lasted for centuries. In the United States alone during the 1860s, over 600,000 soldiers and citizens lost their lives in a bloody Civil War that culminated in abolishing the South's so-called "peculiar institution" of enslavement. In addition to the horrendous loss of American lives, tens of thousands of others from societies around world sacrificed themselves in the fight to end this abusive system.

The above two examples of Nazism and slavery highlight one the major weaknesses of moral relativism and its two forms of ethical subjectivism. If, as moral relativists contend, all individual or group values are equal in moral worth and no broader ethical framework or principles exist to critique the cruel behaviors that stem from these and other evil systems, then there is no way to make the case for values that enhance life over those that destroy it. In the final analysis, when anything goes, everything goes.

This implies that if no moral comparisons can be made, then killing and kindness are on an equal ethical plane. From our perspective it is reasonable to conclude that for the majority of the world's population, with rare exception, any ethical system incapable of separating killing from kindness would be judged as morally inferior to those that do. As we will explain soon, this is in fact the case in all societies without exception.

Moral Relativism and Cultural Pluralism-Moral Universalism

In addition to clarifying our use of the terms ethics, morality, cultural relativism, and moral relativism/ethical subjectivism, we need to make two further refinements. The first entails distinguishing the concepts of moral relativism from cultural pluralism. Although they are often used interchangeably, they are different. As stated, relativism refers to the absence of a universal set of moral standards. Pluralism, on the other hand, holds that while multiple value systems exist, this condition is compatible with the presence in all societies of a universal morality. In addition, pluralism implies that a broad moral framework can be used to judge the ethical worth of different value preferences.

In the above example of slavery, the worldwide rejection of owning and selling people as property would not have occurred without the existence of universal moral principles that virtually every culture on Earth embraces. While different groups have varied in their views of slavery with some wanting to keep it, others to modify it, and still others to abolish it, over time the pluralism of value positions gave way to an underlying moral universalism that led to the worldwide ethical condemnation and legal abolition of this practice. The parallel with Hitler's Third Reich is striking. There is no inherent standard in moral relativist beliefs that would lead to the denunciation of Nazi cruelty.

Moral Universalism and Moral Absolutism

Building on the above definitions, we need to make one final clarification that involves distinguishing moral universalism from moral absolutism. Like others that we described above, these two are also seen as synonyms. However, they are not. As stated, moral universalism refers to values that appear in all societies irrespective of pluralistic differences. This implies that moral universalism is compatible the diversity of religious and philosophical worldviews that appear in different cultures and religions around the world, such as Hinduism, Buddhism, Judaism, Christianity, Islam, agnosticism, atheism, among others.[4]

Moral absolutism is different, because it is based on the belief that only one of the many pluralistic positions is true and that the others are either false or inferior. Examples are numerous, especially in the area of religious commitments where the followers of a particular faith perceive that they and they alone possess the truth or at least more of it than anyone else. When Christians, Muslims, Hindus, Buddhists or any other religious group claims that salvation or enlightenment comes only by accepting their doctrines or specific moral codes, they have crossed the line from pluralism and into narrow absolutism.

Contrary to absolutism, which makes exclusive claims on behalf of only one alternative, it is in the nature of universalism to identify the common elements that appear in all pluralistic positions. Like moral relativists, those who hold to some form of moral absolutism deny that a universal morality can be found within the world's myriad religious and philosophical

4. For a detailed discussion of how the world's religions differ from each other, see McFaul, *The Future of God in the Global Village*.

traditions. While they differ widely in their reasoning, the followers of both relativism and absolutism perceive that universalism and pluralism are irreconcilable. However, we will show that the opposite in true: moral universalism and pluralism coexist compatibly within and across all the diverse cultures of the world.

The Evidence for Moral Universalism

We will make the case for moral universalism at two levels. First, we will expose the logical inconsistencies of moral relativism, and second we will show how the cross-cultural empirical evidence supports moral universalism by detailing how it is compatible with the pluralism of religious, philosophical, and cultural beliefs and practices that exist throughout the Earth's diverse societies.

Logical inconsistencies of Moral Relativism

The most glaring internal weakness of moral relativism involves the claim that we should not pass judgment on the ethical views of others, because all perspectives are equal. However, we claim that this is a contradiction at the most fundamental level. If, as ethical subjectivists assert, all values are exclusive and therefore confined to either the individuals or groups that hold them, then intolerance and tolerance possess identical ethical significance, and no broader perspective exists for differentiating one from the other as a preferred ethical alternative.

Why anyone would favor tolerance over intolerance, or vice versa? The answer is not at all clear unless a more encompassing framework provides the foundation for choosing the former over the latter, that is, for determining that being nonjudgmental or tolerant of differences is ethically more compelling than being judgmental or intolerant. When subjectivists of any stripe argue that moral diversity should be tolerated no matter how much individual or group values might differ from each other, they have left the realm of relativism and entered the world of universalism. How a universal principle of tolerance extends from moral relativism, where both tolerance and intolerance carry equal moral weight, is never explained— merely asserted.

Historical and Political Setting of Moral Relativism

Given this inconsistency, can we say anything in favor of moral relativism? The answer to this question is yes, especially when we understand how and why it emerged as a popular ethical theory throughout the twentieth century. One of the best ways to grasp moral relativism's rise in popularity is to see it as more than just the philosophical left hand—so to speak—to the "right hand" of cultural relativism. While there is little doubt that the combination of ethnography and philosophy led to the theory's allure, this is not the whole story. Of equal importance, although less well known, is moral relativism's connection to history and politics.

The historical and political aspects of moral relativism were a direct offshoot of four centuries of Western colonialism. In a nutshell, while the modern form of moral relativism began as an ethical theory during the early twentieth century, it evolved over the next several decades into a movement of outspoken critics who opposed Europe's domination of non-European nations. Using the descriptive field studies of modern ethnographic science to bolster their belief in the equality of all cultural norms, moral relativists launched a critique of the Euro-American condescension toward other societies, which permeated the colonial era.

They argued that when European countries conquered non-Europeans, they not only became dominant politically and economically, they also imposed an arrogant mind-set of "our way of life is better than yours" on the subjugated but culturally dissimilar masses. In light of their strong belief that modern ethnographic research demonstrated that the norms of all societies (from tribal to industrial) are equal, they were deeply offended by this patronizing sense of superiority. It was out of the historical setting of modern anthropology along with its scientific field studies that they transformed their ethics into politics and aimed their protest at an arrogant Euro-American moral absolutism.

Thus, once we grasp the political as well as the scientific and philosophical dimensions of moral relativism and locate it in a specific historical context, we can understand why and how it became caught up in a logical inconsistency. Even as moral relativists were calling for tolerance toward all intercultural normative differences, they were demonstrating intolerance toward what they perceived was Euro-American cultural arrogance. In sum, the best way to grasp the origins of this contradiction is by viewing it at two different levels—one ethical and the other political.

The Decline of Moral Relativism and Rise of Moral Universalism

Then, as the world transitioned from the twentieth to the early twenty-first century, new patterns began to emerge in conjunction with the demise of the colonial system. As liberation movements pushed for and achieved their political independence in nation after nation during the early to mid-twentieth century, European domination gave way to Asian and African demands for equality and justice. In the process, the attitude of condescension lost ground to new voices that called for an end to exploitation and patronizing practices. As the era of multinational independence emerged around the world, moral universalism began to replace moral relativism, and in the process exposed another of the relativism's weaknesses.

The emergence of movements for national independence within the conquered nations lifted into high relief an unforeseen irony. While Western moral relativists were making the case for tolerance based on the ethnographic studies of Western scholars, non-Westerners in liberation movements began pointing out that the defense of tolerance by itself merely reinforced the unjust status quo that existed within the subjugated nations that were seeking their liberation from European control. As a result, Western critics unexpectedly found themselves subject to criticism from the non-Western world they were defending.[5]

Although tolerance of cultural differences was lauded as a step in the right direction, it did not go far enough. European exploitation of the non-European populace was only half the picture. For many political leaders in the newly forming Asian and African nations, abusive practices existing within their societies also needed to be addressed. With this shift in emphasis, a new door opened on the search for an alternative ethical framework that could address unjust practices at two levels: first, those that Westerners imposed on the non-Western world (colonialism) and, second, those that existed within the non-Western world itself (newly created nations). In a word, an all-encompassing ethical perspective was required to deal more broadly, indeed, globally, with issues that moral relativism by itself could not.

---

5. For an excellent summary of the history of the rise and fall of moral relativism in light of colonialism, see Diamond, *In Search of the Primitive*. According to Diamond, "Relativism is the bad faith of the conqueror," 110.

Moral Universalism and the Impact of New Evidence

As the search of a universal moral perspective moved forward, it gained support from many late twentieth century cultural anthropologists who were conducting new field studies of their own. This younger cohort of scholars fanned out in two directions. First, many of them returned to the small, once isolated tribes that the earlier generation had studied in order to assess the impact that modernization was having on those cultures. In the process, they made discoveries that expanded, and in many cases contradicted, the findings of earlier researchers. Based on cross-societal comparisons, they concluded that cultural differences were not as extensive as earlier studies claimed. In addition, they began to identify a cluster of values that the groups they were studying held in common. This discovery led to the next step of questioning whether these shared values might be present in other societies as well.

We can answer this question by turning to the studies of the second group of anthropologists who were doing field research in cultures that had not yet been explored. As this group began sharing the results of their work, it became increasingly clear that their cumulative findings dovetailed with the discoveries of the first group of anthropologists, who returned from formerly investigated cultures. Gradually, a new paradigm began to emerge. Based on the careful cross-cultural research of both groups, the scientific evidence started tilting away from the moral relativists who claimed it supported their ethical views and toward those who were championing a different perception—one that combines cultural pluralism with moral universalism.

The earlier anthropologists were pathfinders who spearheaded the initial exploration of previously unexamined non-Western cultures. Their discoveries opened the world's eyes to the enormous range of cultural practices that exists across societies. Then, as the emerging field of anthropology began to mature, the next generation of scholars started observing that in the midst of widespread dissimilarities, there is a deeper level of shared values. This, in turn, triggered the search to identify the ethical elements that appear amidst the broad range of cultures as well as how they connect to the variety of practices that separate societies from each other.

Although the limitation of space does not permit a comprehensive review, the following examples highlight some of the key persons, discoveries, and organizations that led to the transformation from relativism to universalism during past several decades. According to well-known sociologist

Wendell Bell, "Today, we know that cultural diversity in the world has been greatly exaggerated. . . . Exemplars that have sustained the belief in cultural diversity and relativism have been found wanting in restudy after restudy."[6]

In addition, Donald E. Brown observes that follow up field research has shown that contrary to Margaret Mead's early studies adolescents in Samoa experience as much stress as Western teenagers and that the American male and female temperaments are identical to and not the reverse of those of the New Guinea people. Also, in contrast to earlier accounts, further research has shown that Hopi Indians do share with Americans similar views of time. No matter the location, people classify colors in similar ways despite diverse dress codes. Facial expressions and body gestures are interpreted in the same way in all cultures. In all societies, parental warmth and love affect children in positive ways.[7]

Cross-cultural similarities are also present at basic bilingual levels despite language dissimilarities. Renowned linguist Noam Chomsky demonstrated in the early 1970s that even though grammar and syntax separate societies from each other, underneath these differences, all persons hold in common a framework of semantic components through which they carve up the world. It is because of this framework that human interaction is possible when any given language is translated into another.[8] Otherwise, cross-cultural communication would be severely limited and in many cases not exist at all.

In support of these research results, several writers began identifying lists of values that they believe people everywhere hold in common. Many examples could be included, but two are noteworthy of mention because they reinforce the results of earlier anthropological evidence. Based on years of cross-cultural comparisons, ethicist Rushworth M. Kidder has set forth the following well known summary of the universal values that cultures around the world embrace. He identifies the ten value areas that the leaders of the societies he studied ranked as among their highest: achievement, benevolence, conformity, enjoyment, power, security, self-direction, stimulation, tradition, and unity.[9]

In a gathering of over 200 delegates from more than one hundred world faiths, the Parliament of the World's Religions met in Chicago in

---

6. Bell, *Foundations of Futures Studies,* 175.

7. Brown, *Human Universals.*

8. Chomsky, *Language and Mind.* Bell, *Foundation of Futures Studies,* 171–77.

9. Kidder, *Shared Values for a Troubled World.*

1993 to celebrate its centennial anniversary since first meeting in 1893. The Parliament's major document, *Toward a Universal Ethic*, highlights an array of universal standards that cut across all religions. The following list represents many of the values that the Parliament endorsed as preferred universal cross-cultural norms: tolerance toward difference, freedom of expression without inflicting harm on others, truthfulness and honesty, integrity, social justice and nonviolent change, kindness, forgiveness, an end to discrimination because of race, sex, age, physical and mental ability, religion, nationality, among others.[10]

Based on the above examples, we can conclude that the weight of evidence from multiple voices in diverse settings during the past several decades has tipped the scale toward universalism and away from relativism. This change did not occur all at once but rather gradually over decades through the cumulative efforts of many scholars, writers, and religious leaders who were working in separate areas. What they shared in common was—and still is—the desire to dig to the deepest level possible to uncover the values that exist beneath the surface of initial observations of how societies differ in their languages, food and dress norms, and other social customs.

Universal Reciprocity Norm—
Ancient and Modern Evidence

The discovery of common patterns and preferences that appear amidst the pluralism of diverse traditions led some observers to ask whether there might be a single ethical standard that goes deeper than all the rest, one that informs all of the similarities as well as differences at whatever level. This resulted in recognizing that a unifying principle can be found at the center of all societies: a norm of Universal Reciprocity. As will be described below, this norm exists in all the great religions and philosophies that are spread across the Earth and in all the societies they have helped create.

The Universal Reciprocity Norm also goes by the name of the Golden Rule. Further fine-tuning of this rule can be made by subdividing it into golden and silver forms of expression, which are based on either positive or negative formulations. The positive form of the Golden Rule is "Do to others as you want them to do to you;" whereas the negative form is "Do

10. Parliament of the World's Religions, *Toward a Global Ethic: an Initial Declaration.*

not do to others as you do not want them to do to you." Table 11 summarizes how the great religions that appear across the spectrum of diverse world cultures express the Universal Reciprocity Norm in both its golden and silver versions.[11]

Table 11. Golden and Silver Versions of the Universal Reciprocity Norm.

The Golden Version

Judaism	Thou shall love thy neighbor as thyself. (*Leviticus* 19:18)
Christianity	All things whatsoever you would that men should do to you, do you even so to them. (*Matthew* 7:12)
Islam	No one of you is a believer until he desires for his brother that which he desires for himself. (*Hadith*)
Jainism	A man should journey treating all creatures as he himself would be treated. (*Sutrakritanga* 1.11.33)
Sikhism	As thou hast deemed thyself, so deem others. (*Guru Granth Sahib*, p. 1299)
Daoism	Regard your neighbor's gain as your own gain and you neighbor's loss as your own loss. (*Tai Shang Kan Ying Pien*, 213).
Shinto	The heart of the person before you is a mirror.
Native American	Respect for all life is the Foundation. (*Great Law of Peace*)
Plato	May I do to others as I would that they should do unto me.
Seneca	Treat Your inferiors as you would be treated by your superiors. (*Epistle* 47:11)
Roman Pagan Religion	The law imprinted on the hearts of all men is to love the members of society as themselves.
Confucianism	Tsi-kung asked, "Is there one word that can serve as a principle of conduct for life?" Confucius replied, "It is the word 'shu'— reciprocity." (*Doctrine of the Mean* 13:3) note: underlining added.

---

11. Granoff, "Peace and Security," 627–30.

The Silver Version

Judaism	What is hateful to you, do not do to your fellow man. That is the law; all the rest is commentary. (*Talmud, Shabbat* 31a)
Buddhism (1)	Hurt not others in ways that you yourself would find hurtful. (*Udana-Varga* 5:18)
Buddhism (2)	A state that is not pleasing or delightful to me, how could I inflict that upon another? (*Samyutta Nikaya* verse 353)
Hinduism	This is the sum of duty; do not unto others which would cause you pain if done to you. (*Mahabharata* 5:1517)
Jainism	Neither does a wise person cause violence to others nor does he make others do so. (*Acarangasutra* 5:101–102)
Zoroastrianism	That nature only is good when it shall not do unto another whatsoever is not good for its own self. (*Dadistan-I-Dinik* 94:5)
Yoruba Wisdom (Nigeria)	One going to take a pointed stick to pinch a baby bird should first try it on himself to feel how it hurts.
Socrates	Do not do to others that which would anger you if others did it to you.
Confucianism	One should not behave toward others in a way which is disagreeable to oneself. (*Mencius* Vii.A.4)
Confucius	Do not unto others what you would not have them do unto you. (*Analects* 15:23)

A thoughtful examination of this list reveals that the two versions of the Universal Reciprocity Norm or Golden Rule are two sides of the same coin. The golden version of do good is one side and the silver version of do no harm is the other. In Kent M. Keith's view, these two sides can be refined even further. Doing no harm refers to: do not lie, steal, cheat, accuse others falsely, commit adultery or incest, abuse others verbally or physically, or destroy the natural environment on which life depends. Doing good means to be: honest and fair, generous and faithful to family and friends, and kind to strangers. It also means to take care of children when they are young and parents when they are old or cannot care for themselves, to respect all life, and to protect the environment on which all life depends.[12]

12. Keith, *Morality and Morale.*

Immanuel Kant (1724–1802 CE), who is arguably the most important Western philosopher during the past 250 years, wrote extensively on the golden and silver versions of the Universal Reciprocity Norm or Golden Rule that he called the Categorical Imperative. He elaborated the norm into different formulations and claimed that it serves as the foundation of all morality that he called Practical Philosophy. In his view everyone has a perfect or imperative duty to do no harm (the silver version). Over and above the no harm principle stands everyone's imperfect duty, which is based on doing the good (the golden version) voluntarily out of personal benevolence and generosity.[13]

Whether stated as the silver or golden version, the Categorical Imperative in both its perfect and imperfect interpretations, or as refined and expanded into its various elements, the cross-cultural nature of the Universal Reciprocity Norm is readily apparent in the above examples. What is truly remarkable is that this norm emerged centuries ago when the world's cultures were isolated from each other by high mountains and wide oceans.

Anthropologists call this a cultural parallel, which means that the same norm or pattern arises in different parts of the world such as West Asia/Middle East (Judaism), South Asia/India (Hinduism), Central Asia/China (Confucianism), Africa/Nigeria (Yoruba), and North America (Native American tribes), without prior intercultural contact or the transmission of ideas from one society to another. It is from this perspective that the Reciprocity Norm qualifies as a moral universal that appeared many millennia ago, spread steadily throughout the world's diverse cultures, and continues down to the present day.

Nowhere else is this more evident than in the work of the United Nations since the end of World War II. As the world has evolved during the past several decades in the direction of becoming a global village,[14] the UN's 1948 Universal Declaration of Human Rights provides some of the strongest evidence to date that a ubiquitous moral principle and its extensive interpretations and applications across cultures continues to have a profound impact on the Earth's diverse nations. Endorsed by virtually all

13. Kant, *The Critique of Practical Reason*. Kant, *Fundamental Principles of the Metaphysics of Ethics*. Kant's first version of the Categorical Imperative is that we should always act in such a way that the maxim or motive of our actions can be universalized and reversed back to ourselves. His second version states that we should treat persons as ends in themselves and never merely as a means to an end.

14. For a description of the global forces that are helping create the global village, see McFaul, *The Future of Peace and Justice in The Global Village*, 3–14.

countries, the Declaration defines the reciprocal expectations that people on all continents have embraced ethically, politically, and legally.

The Preamble and Thirty Articles set forth the following universal rights and freedoms: life, liberty, security, removal of all forms of prejudice based on race, religion, sex, age or other demographic factors that have nothing to do with an individual's dignity or innate capacities, elimination of torture, inhumane punishment, and arbitrary arrest and retention.

The Declaration also supports the right of privacy in home, family, and the right to own property, freedom of geographical movement, the right to marry and to equal rights in marriage, association, the right to democratic participation in government based on public will, to work with equal pay, universal education, participation in science and the arts, and to a peaceful international order.

Since the initial writing and approval of the Declaration, the UN has also drafted a series of expanded resolutions aimed a specific populations such women, children, indigenous populations, AIDS victims, among others. In the year 2000, the UN General Assembly adopted a Millennium Declaration in which it re-affirmed many of its previous pronouncements related to values and principles by which the world should be governed, including freedom, equality, solidarity, tolerance, respect for nature, shared responsibility, eradication of poverty, protection of vulnerable populations, and universal justice.

When viewed from the perspective of its many declarations and documents, the UN for more than six decades has fleshed out how all-encompassing ethical standards should be applied to the conduct of daily life everywhere. These range from the rights of individuals to the freedoms that should be included within the legal frameworks that govern entire communities throughout the emerging global village. In keeping with the theme of this chapter, another way to say this is that the world's single most inclusive organization in which all of the world's nations hold membership has articulated through its all-encompassing public statements the principles of doing no harm (the silver version of the Golden Rule) as well as doing the good (the golden version).

Furthermore, as is readily apparent, the idea of justice, as discussed in the last chapter, is one of the core standards that the UN's and other ethical and legal declarations include. The reason for this is clear. The aspiration of people everywhere to improve the conditions of justice in their societies is one of the most important values that cuts across all cultures and is embedded within the all-encompassing Universal Reciprocity Norm.

The Evolution of Values and the Universal Reciprocity Norm

Before concluding this chapter, we need to address one more matter. In the course of our discussion, we observed that during the past 5,000 years, humanity has evolved from a checkerboard pattern of many small and isolated communities to its current condition of large nations and regions that are traveling together down the road toward greater global integration. Since the end of WWII, the world has been witnessing the steady erosion of support for moral relativism and increasing acceptance of moral universalism coupled with pluralism.

Given that the ethical evolution from narrow to broad standards appears to parallel the social transition from smaller to larger groups, does this imply that past practices such as slavery should not be judged as immoral until such time as society declared them to be so? Stated differently, can present voices assert that former practices were immoral even though they were not considered to be so when they prevailed in the past?

From the perspective of the Universal Reciprocity Norm and the justice value that has existed in dissimilar cultures for thousands of years, we can assert that the slavery system has always been immoral even though past societies practiced it for centuries. It is not as though the world created a new reciprocity standard of justice after slavery came into existence in order to abolish it. Rather, as moral norms developed over time, the innate but latent Universal Reciprocity Norm that sits at the center of all societies gradually aroused the individual and collective conscience to, first, critique this cruel system and, second, to legally eliminate it. The same can be said of Nazism during the 1930s and 1940s. Nazi practices were immoral and unjust not only because non-Nazi societies declared them to be, although they did. They were immoral and unjust because humanity said so and with rare exception still does.

While we would not say that eradicating slavery and Nazism should not be viewed as moral progress, we also would not conclude that the progressive elimination of these and other expressions of cruelty resulted from the emergence of entirely new ethical imperatives that had no prior existence. The golden (do to others) and silver (do not do to others) versions of the Universal Reciprocity Norm have been embedded within the human mind and heart for thousands of years as revealed in parallel texts that first emerged on different continents in the ancient world.

It seems nothing short of paradoxical for us to say that morality has moved forward by going backwards—so to speak—and that what seem to be new values are merely updated and continuing applications of the world's oldest value—the Universal Reciprocity Norm. Another way to say this is that given the hard evidence that stretches from the ancient past to the modern present, we can reasonably conclude that for centuries humanity has been structured for a universal morality.

Conclusion

Before turning to the final chapter, we can draw a number of conclusions based on the above discussion. We begin with a general summary of the terminology we used throughout this chapter.

- Ethics and morality: used interchangeably to designate the quest for common values.

- Moral relativism: the belief that common values do not exist.

- Ethical subjectivism: two types of moral relativism—individual and group.

- Moral universalism: the belief that common values do exist.

- Science of Ethnography: Descriptive study of cross-cultural similarities and differences.

- Cultural relativism: Ethnographic research on cultural differences that moral relativists used to support their position.

- Pluralism: multiple cultural practices—compatible with moral universalism.

- Moral absolutism: the belief that only one point of view possesses the truth and all others are false or inferior—like moral relativism it is incompatible with moral universalism.

- Universal Reciprocity Norm: shared by all societies since the start of civilization.

- Golden and silver versions of the Universal Reciprocity Norm or Golden Rule: do to others and do not do to others.

Through the above discussion, we have shown that empirical evidence supports the case that a universal morality has always existed at the heart of all societies from the beginning of human evolution thousands of years ago despite the popularity of moral relativism during much of the twentieth century. Early ethnographers in the field of anthropology provided moral relativists with scientific descriptions that they believed supported their claim that universal moral standards do not exist. However, follow up studies from a variety of settings along with the decline of colonialism revealed that a deeper core of shared values lies at the center of diverse cultures despite the many pluralistic practices that separate them from each other.

This triggered a continuing search at the deepest level possible in a variety of societies for a possible unifying cross-cultural ethical principle. This resulted in recognizing that cultures the world over share a common value called the Universal Reciprocity Norm. Ancient religious and philosophical writings reveal that this norm emerged early in the development of different civilizations that arose in isolation from each other. The two formulations are the positive or golden version of do to others as you want them to do to you and the negative or silver version of do not do to others what you do not want them to do to you. Virtually every past and present society incorporates one or both versions of the Universal Reciprocity Norm or Golden Rule.

Despite the existence of cruel and immoral practices that have existed for centuries, and still do, in one society after another, the world has made steady progress in applying the norm to an increasing number of human activities, such as the expanding justice for all by eliminating slavery, ending colonialism, spreading democracy, and by extending rights and freedoms to formerly excluded populations such as minorities, women, children, and other vulnerable populations. The United Nations' numerous Human Rights Declarations and the Parliament of the World's Religions' advocacy of a Universal Ethic represent the pinnacle expressions of this norm as witnessed by near universal support from government leaders and influential citizens around the world.

Thus, as we bring this chapter to an end and move toward the next and final chapter, we believe that the above cumulative evidence points in one direction and one direction only: humanity is structured for a universal morality.

8

God Is Here to Stay

Introduction

In this final chapter, our goal is to first review and then to integrate the available and relevant scientific knowledge that we presented in the previous chapters in order to establish confidence levels related to belief or disbelief that an intelligent designer called God created the universe. As stated earlier in the Introduction, we believe that the cosmos is theologically ambiguous as demonstrated in our review of the Cosmological, Teleological, Moral, and Ontological arguments. While evidence can point in one direction or the other, the core question comes down to this: based on our current level of knowledge, is it more reasonable to believe than not to believe that God exists?

To start, in order to avoid confusion, we need to clarify our method and underlying assumptions. We recognize that like all humans we are vulnerable to filtering the facts and applying our own worldview through preexisting mindsets and that all of us impose our personal preferences on how to interpret the data. As we stated in the Introduction, we are keenly aware that the relationship between deductive and inductive reasoning is subtle and complex and that it is not always easy to determine where one leaves off and the other begins. At the same time, we are deeply committed to searching for the truth through the rational/empirical methods of modern science that involve a disciplined and systematic approach to the study of nature and society. As a result, in order to keep our biases at bay to the fullest extent possible, we base our confidence levels on the best available scientific knowledge.

L-M Scale

Before turning to the evidence for this connection, we need to discuss our procedure for determining different confidence levels. For the purposes of this book, we have developed a confidence scale that appears in Figure 15 below.

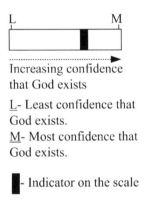

Increasing confidence
that God exists

L- Least confidence that
God exists.
M- Most confidence that
God exists.

- Indicator on the scale

Figure 15. Confidence scale.

It is based on an L to M continuum, where L on the left indicates the least amount of confidence that God exists and where M on the right designates the most. The indicator bar can be moved back and forth on a sliding scale from L (least confidence) to M (most confidence) depending on our perceived level of support. The position of the indicator represents our conclusions based on the best, current, and relevant information.

In addition, we do not consider that the confidence levels that we assign at the end of each section stand apart from those of other sections. In the final analysis, it is the combination of all summaries that provides the most comprehensive level of confidence. We consider that each summary is part of an integrated whole that translates into the strongest possible conclusion that we will make at the end of the chapter. We are now ready to turn to a brief summary of the main evidence we presented in each of the previous chapters starting with chapter 2 on knowledge.

Establishing Confidence Levels

In this section of chapter 8, we will summarize the main issues that we have presented in chapters 2 through 7. We will also establish confidence levels for each separate area as well as for all areas combined, which we will do at the conclusion of this chapter. We begin with chapter 2, which centered on knowledge.

Knowledge

We begin the task of summarizing the main points of the preceding chapters with the topic of knowledge because without this foundation there is nothing else that we or anyone else can say. After our Introduction in chapter 1, we turned immediately to the topic of knowledge in chapter 2—and for good reason. Our objective was to set a broad framework of knowledge within which we would then be able to examine in greater detail the several topics that we included in chapters 3 through 7. We noted that not only has humanity expanded the total pool of knowledge, that is, the quantity or amount of it, but the quality and accuracy of it as well. This trend is particularly noticeable since the rise of modern science 200 years ago, which has yielded a level of knowledge that pre-moderns could barely begin to imagine.

Another way to say this is that we humans were not born with a mental database containing all there is to know saved in our brains. Instead, as part of our human advancement, like the universe as a whole, our expanding reservoir of knowledge—in terms of both quantity and quality—is evolutionary in nature. The evidence we cited in chapter 2 makes this abundantly clear. In the area of communication, we have evolved from limited face to face contact to sending correspondence of all kinds at the speed of light without geographical constraints. Whereas walking was once our primary mode of transportation, we now fly through near-Earth atmosphere in jet aircraft and into outer space in rocket driven spacecraft. We explore our solar system with moon and Mars rover landings and planetary flybys.

The list continues as the evolution of knowledge has expanded into countless other areas as well. When our ancient ancestors gazed upward at constellations in the starry sky, little did they know that one day observational cosmologists would calculate that the universe started 13.7 billion years ago in a phenomenal burst of energy we call the Inflation or Big Bang.

Nor could they even have begun to contemplate the atomic structures that lie hidden beneath everyday objects but which we now know exist because of the evolution of modern physics, which constantly stretches its horizons into new realms of research.

Furthermore, in keeping pace with these extraordinary achievements, progress in modern medicine has evolved through advances of its own starting with discoveries that yielded an accurate understanding of human anatomy and its functions. The rapid changes that occurred from this point forward branched out in many directions ranging from the development of vaccines and antibiotics to performing open heart surgery and organ transplants. Cloning and stem cell research followed close behind. In addition to increasing average life expectancy, decoding DNA and mapping the human genome opened up for the first time in the human evolution the potential to eradicate inherited genetic diseases and to extend the maximum human life span beyond 120 years.

While many more areas of research and discovery could be added to this brief summary of the impact that scientific knowledge has had on human advancement, it goes without saying that none of this would have been possible without the evolution of higher brain functions that are connected to the cerebral cortex. As this newer part of the brain evolved, so did the human capacity to create, retain, and recall knowledge that is the essential prerequisite for our ongoing development as a species. As discussed earlier, we recognize that thinking as a mental event is correlated with the biochemical processes that occur in the brain as a physical object. We do not presume to comprehend this correlation in terms of causality. Despite competing interpretations, the brain-mind-knowledge connection remains one of the unsolved mysteries of human experience. However, irrespective of our lack of understanding on this issue, it is clear is that unlike other species that survive by instinct we humans have evolved on Earth by creating linguistically based knowledge as our primary means of survival and advancement.

One final question remains to be answered before we proceed to the next section. Do advancements in both the quantity and quality of scientific knowledge during the past 200 years, as we have described them in Chapters 2 through 7, increase or decrease our confidence in the existence of God? This question is not easy to answer because by itself the evolution of knowledge as humanity's primary means of earthly adaptation based on the inexplicable but highly correlated connection between the cerebral cortex and abstract thought can be viewed from both naturalistic and the-istic perspectives.

By itself, the existence of knowledge is soft evidence in making the case for God's existence because non-theists can view the evolution of knowledge as humanity's means of adaptation on Earth without the need of a God hypothesis to explain it. However, when coupled with the other issues we discuss in the subsequent sections of this chapter, it plays a central role in tipping the scale in the direction of belief in God, especially in light of the strong empirical and mathematical approach we have employed throughout this book. For this reason, we are 60 percent confident that the best available scientific knowledge supports the conclusion that an intelligent designer called God created the universe.

We understand that there is a personal or subjective aspect to setting confidence levels in this and the other areas we describe below. We also recognize that different persons will set dissimilar confidence levels depending on how they interpret the scientific evidence. We do not see this as a problem. Given our perception that the universe is theologically ambiguous, it could not be otherwise. What is essential to our approach is that our confidence levels are built on a solid foundation of the best scientific evidence available as we summarize it in the following sections.

Next we turn to a brief review of the key issues described in chapter 3 on the evolution of the universe.

The Universe Is Structured for Conscious, Self-Aware Life

In this section we summarize the key conclusions related to the universe origin of the universe and of the evolution of conscious, self-aware life on Earth as detailed in chapter 3. As in the case of the preceding section on knowledge, we will set our confidence level after reviewing all of the relevant evidence.

Laws of nature and their tight limits allow for conscious, self-aware life: Chapter 3 focuses on the Anthropic Principle that starts by recognizing that the basic laws of nature governing our universe have numbers (constants) that are embedded in them. If these numbers deviated even by small amounts, then life as we know it would not have developed in our universe. These laws span a range from very small, such as subatomic particles and their forces, to very large, such as the origin of the universe in the Big Bang and its subsequent development. From the perspective of the best knowledge available, there do not appear to be any evidences that refute the Anthropic Principle.

It is not currently known why this set of numbers appears in the basic laws of nature that govern our universe. Despite this uncertainty, what is clear is that the cosmos is structured in such a way that it evolved into the formation of carbon based and water dependent life on our planet 3.85 billion years ago and ultimately to the emergence of conscious, self-aware humans 200,000 years ago. We have no commonly accepted standard model by which to explain this process. However, based on current scientific evidence, since the Big Bang, evolution has led to the creation of life on Earth as we understand it according to the narrow numbers associated with the Anthropic Principle.

Gravitational constant has a numerical value within tight limits for life: Gravity is a special case of the Anthropic Principle. Gravity is a force that we feel all the time. It extends across the universe and affects entire solar systems, black holes, and galactic clusters. If gravity varied even slightly, then the original dust and rocks that shaped our solar system would have accumulated differently to produce a different set of planets and their orbits. The Earth (if there was such a planet) would likely have formed at a different distance from the Sun, which itself would probably be burning differently.

In addition, with slightly altered gravitational forces, the Earth would have different characteristics that would not be so life friendly. For life to originate on Earth, or precursors to be transported to Earth, our planet would have needed a receptive environment for germination to start. The atmosphere would have needed the right ranges in temperature, chemical composition, and substrates and/or solvents for those original organic reactions to occur. Additional requirements are needed such as an orbit stabilizer like the Moon, a magnetic field to protect life from the Sun's solar wind, and an extensive amount of time, all of which are gravity dependent.

Based on current evidence, the strength of Earth's gravity must be within a narrow window. At the same time, we recognize that there are likely other effects that can be neither known nor explained. Taking all this into account, the allowable window in the gravitational constant is estimated to be plus or minus 1 percent; and these narrow numbers are consistent with our understanding of the Anthropic Principle.

Matter-Antimatter imbalance: It appears that at the beginning of our universe equal amounts of matter and antimatter were produced. Yet all current evidence indicates that the universe is only made of matter and not antimatter. We do not see isolated regions of space where antimatter exists and is separated from matter. It is currently not known why matter dominates antimatter.

The imbalance between matter and antimatter is atypical because most laws of nature involve balances. If there was a perfect balance, there likely would have been complete annihilation of all matter no possibility of conscious, self-aware life. Once again, we find that this narrow imbalance gives additional credence to the Anthropic Principle.

Nineteen Integrated Characteristics: See chapter 3, Table 1 where nineteen characteristics are listed for consideration of how they contribute to the evolution of the universe and conscious, self-aware life. Other characteristics could have been added. Each characteristic is segmented into three results, two of which are not life friendly and one that is. The basic idea for Table 1 is to consider all nineteen together as an integrated whole and not to view each of them as a separate item.

Well known and accepted laws of probability were used. Generous assumptions were made, favorable for a random and undirected universe. Even with this unbiased approach we conclude that the numerical chance for life to emerge randomly is incomprehensively small. The key was the consideration of all nineteen effects as a whole. This appears to greatly increase our confidence that from the time of the Big Bang the universe in general and conscious, self-aware life in particular evolved through a process of design rather than randomness.

Predictability arising from natural law: The Standard Model for subatomic particles and forces has been almost universally accepted. It has been extolled as one of the great scientific triumphs of the twentieth century. The main reason it has garnered such lofty praise it because researchers have used it repeatedly to make recent predictions, confirmed with measurements. It has been tested experimentally many times, in many ways, and in many diverse laboratory settings. On each occasion it has proven to be successful.

In another example of predictability, in 1928 the theoretical physicist Paul Dirac correctly anticipated the existence of antimatter before it was known to exist, based only on mathematical calculations. Four years later in 1932 antimatter was discovered in a laboratory experiment, just as he predicted.

The overwhelming successes in predicting how nature behaves leave us with one overarching question: why should this be so? The answer is clear. Despite the extraordinary complexities that we associate with evolution since the Big Bang, the universe is not randomly chaotic. Instead, the narrow physical and mathematical constants by which the cosmos operates indicate that it highly structured.

Furthermore, it is widely accepted that the Big Bang was the singular event that created the universe. There do not appear to be multiple Big Bangs that are now or were once scattered throughout time and space. If there were such multiple starting events, we might expect different types of matter and energy, different laws of nature, and different levels of predictability. If the products of these multiple starting events were to co-mingle, then we might expect various types of boundaries to develop, or various levels of annihilation at the boundaries, and unpredictable types of behaviors due to the different laws of nature. The fact is that we do not observe any of this.

Thus, when all is said and done, the overwhelming quantity and quality of accumulated scientific evidence underpins our perception that only one universe exists, which in turn reinforces the Anthropic Principle. This implies that the natural predictability that we see and experience did not result from randomness. Instead, it increases our confidence that designed structures exist at the core of the cosmos.

The debate over design or randomness in the origin of the universe: Before turning to the next issue of how life emerged from non-life, we conclude this section by discussing some of the alternative views related to the question of whether the universe arose from random processes and forces or by design. Table 12 highlights four major positions that contain substantial disagreements on how to answer this question.[1] For each of the four alternatives, we provide both a Response and Comment. In the discussion that follows Table 12 we compare the strengths and weaknesses of each position.

1. Collins, *The Language of God*, 74–76.

Table 12. Four competing explanations for number of universes and whether or not a designer is needed.

	Response	Comment
1	There is only one universe. This is it. The precise tuning of all the physical constants and laws are required.	A designer is needed.
2	There are other universes. They occur simultaneously with ours or in some time sequence. They may have different physical laws.	No designer is needed. Origin of the other universes is due only to randomness and undirected processes. This is called the "multiverse" hypothesis.
3	There is only one universe. This is it. It just happened to have all the right characteristics for life.	No designer is needed. Origin due only to randomness & undirected processes.
4	There are as yet undiscovered reasons why fine tuning is highly unlikely.	No designer is needed. Origin due only to randomness and undirected processes. We have little understanding of the universe.

The Response summary for Position number 1 is consistent with the accumulated scientific knowledge that we have presented in chapter 3. It assumes there is only one universe that is precisely with physical constants and laws. The most important competitor for number 1 is number 2. If there are many other universes, then of course, we would likely live in the one that is most life friendly. However, response number 2 is far less efficient than number 1 for the following reason. Historically, when there are multiple competing explanations why a given phenomenon exists, science has preferred the simpler one. Number 1 fits this scientific criterion more than number 2.

In addition, there is another important scientific fact that decreases the credibility of number 2. In 2003 and 2004, observational cosmologists used the Hubble Space Telescope to collect light over many hours by examining a particularly faint patch of sky that they thought was more sparsely populated with stars and galaxies than other areas. Much to their surprise, they observed thousands of new galaxies and stars, both nearby and distant, they did not know existed.[2] As a result of these and subsequent

2. "Hubble Goes to the eXtreme to Assemble Farthest-Ever View of the Universe."

observations, we have increased our awareness of the expanding vastness of the universe and the billions of stellar constellations that comprise it. However, despite the advancements of knowledge that cosmologists continue to accumulate, they have yet to detect the existence of multiple universes or signs that they might have existed in the past.

Although hard observational data does not exist, there are some scientists who hypothesize that our current understanding of quantum mechanics and string theory (which has not been commonly accepted) points toward the possible existence of additional universes. Like the lack of evidence cited above for position number 2, positing the possible existence of other universes based on quantum mechanics and string theory is speculative at best. There is no solid empirical evidence to support it. Unlike the Standard Model that we described above, which supports predictability in nature, no such Model exists for either demonstrating the existence of multiple universes or for developing a set of testable observations to show they do. Thus, on the basis of the best scientific knowledge available, at this point in time we are confident that there is only one universe as summarized in response number 1.

In the case of position number 3, we estimate that the probability is extremely small, if not zero, that our universe evolved through random processes. In order for number 3 to have more credibility than number 1, it would be necessary to show how resorting to randomness rather than positing an intelligent designer is a superior way to explain 1) how the universe was created, 2) came to operate within narrow numerical constants (as described in terms of the Anthropic Principle), and 3) evolved in the direction of conscious, self-aware life on Earth. Our confidence is very small that there is an explanation that accounts for these complexities in terms of random processes.

This takes us to the last response box number 4, which strikes us as the most speculative, even more so than number 2, which posits the possible existence of multiple universes. The Comment box in number 4 indicates that we have little understanding of the universe. Actually, the contrary is true. We have a substantial amount of knowledge about how the universe operates even though there is so much more that we have yet to learn. Furthermore, although there are still many uncertainties in our knowledge of how the universe became so finely tuned, this is no reason to conclude that the cosmos came into existence and evolved to its present state as a result of being propelled forward by random forces.

After examining all of the four positions outlined in Table 12, we are left with one question. Which one is the most credible? Based on the fact that the Anthropic Principle is confirmed by the best available scientific evidence, we believe that response number 1 is the most convincing. In addition, given that the universe and conscious, self-aware life would most likely not have evolved as it has outside the narrow boundaries of nature's mathematical constants, we conclude at the 95 percent confidence level that an intelligent designer called God created the universe and has been guiding its evolution since the moment of the Big Bang.

From Non-Life to Life

In this section we summarize the main findings that are described in detail in chapter 4 on how dormant non-living matter became alive and started to populate our planet about 3.85 billion years ago. Some observers have called this the number one unresolved challenge in science today. A successful explanation must include detailed and diverse considerations from the fields of astronomy, geology, Earth science, physics, chemistry, and other fields. This is extremely challenging, maybe impossible.

Non-life to life molecules: When all is said and done, there are two possible explanations for how non-life gave rise to life. The first focuses on extra-terrestrial causes and the second on terrestrial conditions. If we say the origin of life involves extra-terrestrial sources or that the starting molecules came from off the planet, this really does not completely explain how non-life became life. Instead, this explanation merely pushes the sources of the starting materials somewhere away from the Earth. Given this hypothesis, it would be virtually impossible to duplicate in a laboratory setting the conditions that produced carbon based life as we know it.

In order to answer the question of how non-life became transformed into life, we would need to simulate the conditions on Earth as they existed 3.85 billion years ago and test all the promising explanations. This would involve identifying all of the known elements that are related to the modern science of chemistry, physics, biology, thermodynamics, probability, and statistics. From this we would want to develop some kind of standard model for explaining all of the significant data in a way that would have predictive value. The associated results from this model would then be subjected to rigorous scientific scrutiny and testing by independent investigators.

So why is this unlikely to be possible? There are at least three reasons: 1) we do not have a complete idea of what conditions on the Earth 3.85 billion years ago were like when life first came into existence on the surface of the Earth. This means that our confirmation experiments that must account for these conditions will be subject to major uncertainties. There may be ways to do this but they would be very expensive, time consuming, and have large uncertainties. 2) We do not have adequate knowledge about the chemistry and biology of what those earliest life forms must have been like. 3) We lack understanding of how stable and sensitive those earliest life forms were to the chemical and thermal environment and the related changes that occurred over time.

Thus, we are left with a currently irresolvable dilemma. We can speculate that there might have been an infinitely large number of various types of molecules that would be evolving through a process of assembling and disassembling and of folding and unfolding. This would have caused them to arrange and rearrange themselves into various configurations, possibly in different locations, according to the laws of chemistry, probability and statistics, and thermodynamics. While it might be possible that one of these molecular assemblies became transformed to the first life form, on the basis of the best available scientific evidence, at this point in time we do not know for certain.

At the same time, the evidence points in the direction that life started on our planet almost immediately after the Earth could support it. Once life began, it evolved slowly through a process that unfolded within the narrow mathematical constants that are associated with the Anthropic Principle. Eventually, as the evolution of life continued over time, it took an additional 3.85 billion years later before conscious, self-aware human life appeared on Earth.

Earth's environment when life got started: The story of life's evolution on Earth does not stop here. To demonstrate just how extraordinary it was, there is every reason to believe that it should not have happened as it did or that it could have taken an entirely different direction. For example, when the Sun and the solar system were forming and stabilizing the Earth took a physical pounding from asteroids, small planets, and other rocky bodies. During this time it is highly unlikely that life forms, pre-life forms, or molecules could have survived such violent conditions.

As the Sun started to stabilize and as the solar system developed into its current form such devastation on the Earth subsided. A more life friendly environment started to evolve on the Earth's surface. This included the right temperature ranges, atmosphere, abundance of liquid water, radiation from the Sun, absence of most widespread sterilizing events on the planet, and a reduced amount of solar wind reaching the Earth's surface. What most likely happened is that as soon as the Earth became life friendly, life got started. Thereafter, it took nearly 4 billion years for conscious, self-aware life to evolve. In other words, at the right time and in the right incubator, self-conscious, critical thinking human beings appeared on Earth.

Proteins: This complex process of evolution appears even more amazing when we consider the role that proteins play in the formation of biological life. Nearly all chemical reactions in the human body are encouraged by enzymes (proteins). Connective tissue such as hair and muscle are made of proteins. Specific amino acids linked together in a specific order make-up a specific protein. For each type of protein to function correctly in the human body certain conditions must be met such as the electrical charge distribution, the folding of the protein, the sequence of amino acids, and the location of the reaction sites, temperature, and confinement vessel (cell).

When life started on Earth nearly 4 billion years ago its developmental pathway pointed in the direction of this intricate protein nano-machinery (Nano refers to very small) that would eventually support human life. In addition, when we add into this tightly integrated network of proteins the nineteen additional considerations that we discussed in chapter 3 and in the previous section, the account of how the process of evolution produced conscious, self-aware human beings is even more astounding.

This takes us to our final question. Is it more credible to believe that the evolutionary process that produced conscious, self-aware life on Earth is the work of random forces or of an intelligent designer? For us the answer seems clear. When all of the complexities and narrow numerical constants that are linked to evolution since the time of the Big Bang 13.7 billion years ago, it is far more credible to believe that this process did not evolve through undirected randomness but rather according to the design of an intelligent creator called God. Based on this conclusion, we set our confidence level at 90 percent.

The Brain and Spiritual Experiences

Next, we review the main discussion points we presented in chapter 5 in dealing with the issue of human spirituality and the brain.

Long term persistence of religion: We noted at the start of chapter 5 that starting in the seventeenth century predictions regarding the demise of religion have turned out to be completely false. Those who made such predictions like Voltaire assumed that as science advanced, religion would decline. The empirical evidence shows the opposite: The religions of the world are as robust now as ever, and many are growing. In addition, during the past 200 years, several new religions have emerged precisely during the time span when many predicted that science would eradicate religion, which has not happened. Instead, in the midst of this long term trend, both the practice of science and belief in God continue to coexist together in the modern world.

The tension between the two spheres is, of course, not fully reconciled as evidenced by the ongoing debates between creationists and evolutionists and between theists and new atheists. As we stated earlier, we believe that attempts to prove or disprove the existence of God based on scientific evidence are no longer productive or that in order to believe in God one has to reject science and evolution or vice versa. In moving beyond this irreconcilable logjam, we believe strongly that a new epoch has opened up in which scientific knowledge provides increasing support for belief in an intelligent designer called God in comparison to a naturalistic interpretation of the origin and evolution of the universe based on randomness.

We have advanced this viewpoint by showing in chapter 5 how the evolution of the cerebral cortex is tied to human spirituality and the persistence of religion. This newer part of the brain has made conscious, self-aware life possible as modern humans have evolved on Earth during the past 200,000 years. As described in chapter 2 on knowledge, the cerebral cortex gives us the capacity to develop practical knowledge in fields such as communication and the medical sciences that we need for survival and advancement.

It also provides us with the drive to express our inner emotions through artistic expressions such as painting, music, and performances. In the realm of religion, it enables us to ponder the big questions of life such as how the universe began, how it functions, where it is headed, what role we humans play in it, what our destiny might be, and so on. We seek answers to these big questions as much as we endeavor to develop newer forms of practical knowledge.

It is precisely at this point that we believe science, evolution, and belief in God become mutually reinforcing. In chapter 5, we presented research that reveals how children as well as fraternal and genetically identical twins develop their concepts of God. We cited other studies that show how individuals retain their religious beliefs throughout their entire lives, and how religion relates to near death experiences. Once again, we return to the Anthropic Principle because it encapsulates the narrow boundaries within which the universe evolved since the Big Bang. It was not until the newer part of the brain developed the capacity for conscious, self-aware life some 200,000 years ago that we humans who possess it have been able to not only develop many different kinds of knowledge, including the best scientific data available, but to combine it with various forms of self-expression, including art, music, and human spirituality.

Did all of this happen by chance alone through random forces and process? While this is possible, we believe that given the complexity of the universe and the extraordinary intricacy of the process through which it developed an explanation based on randomness inspires far less confidence than the alternative that the Big Bang and evolution are the handiwork of an intelligent designer called God. We end by placing our confidence level for this section at 80 percent.

Humanity and Justice

In our next section we will focus on the issue of justice as described in chapter 6. When modern human first evolved, they joined together in wandering groups of hunter/gatherers whose survival depended on collective self-defense and the ability to obtain food from wild plants and animals. Starting around 10,000 years ago and possibly earlier, small but stable agricultural communities began to appear in the Fertile Crescent of Western Asia, Egypt, India, China, and other regions around the world. The ability to dwell permanently in a single location marked one of the most significant milestones in human evolution. It is commonly called the Agricultural Revolution. The transition from hunter/gatherer clans to settled agrarian communities occurred simultaneously with the successful domestication of both plants and animals.

As these communities remained in one location over multiple generations, they began to develop more complex social structures that included several stratification levels ranging from the lowliest land laborers to the

highest ranking political leaders. The evolution of this increasingly complex arrangement enabled a leisure class to emerge. The people who comprised this class neither tilled the land nor struggled for political control.
Instead, they performed their society's religious functions and engaged in
philosophical reflection.

During the hunter/gatherer stage of human evolution, a clan's para-
mount preoccupation centered on the struggle to survive, which left pre-
cious little time for other activities. With the expansion of permanent
agricultural settlements, the search for ways to guarantee internal harmony
and long term stability grew in importance. It is at this juncture that the
issue of how to fairly distribute society's burdens and benefits began to push
to the surface.

It was in ancient Greece that some of the most significant reflections
on the nature of justice began to appear. More than anyone else, it was
Aristotle (384–322 BCE) who set down the initial idea that later writers
developed in greater detail. Aristotle's original contribution lies in defining
the fundamental principle of justice in the following terms: Justice consists
of treating similar cases in a similar manner and dissimilar cases in a dissimilar manner. Or stated differently, a society is just when those who are
equal are treated equally, and those who are unequal are treated unequally.

From this ancient starting point, subsequent writers elaborated this
foundation concept by adding four universal allocation principles that all
societies apportion differently as they seek to satisfy the fundamental principle. As described in chapter 6, these four consist of equal shares, effort,
achievement or contribution, and need. Advocates for the general theory of
justice hold that the four allocation principles appear in all societies even
though each society differs in the amount of emphasis it gives to each one
of them. There is no "one size fits all" proportioning of these four principles
for all societies. Instead, every society weighs differently the relative value it
attaches to each. For example, some societies like the US place more importance on individual achievement than on the other three principles, while
other societies give greater emphasis to the principle of need.

The effort to identify the justice principles that apply universally to all
cultures does not stop here. During the latter part of the twentieth century
the philosopher John Rawls combined the elements of the general theory
into a systematic approach that resulted in the development of a fully integrated modern social contract theory of justice. Building on the ideas of his
predecessors, he starts by assuming that social stability is related directly to

the degree to which citizens in any given culture perceive that their society distributes its burdens and benefits fairly. As societies become more just they become more stable. When they become more unjust, they become less stable.

For Rawls, two principles lie at the heart of this process. They are the liberty and difference principles. As a society expands it liberties, it becomes more just—and thus more stable. Expanding voting rights to once excluded groups is a case in point. At the same time, societies do not accord everyone equal treatment. Over the course of their development, all societies become stratified into different levels of wealth, power, and prestige. Stratification is an inevitable outcome of social evolution. Differences will exist, and some people will benefit more than others.

The only question is whether the differences are just. According to Rawls, if any given society as a whole benefits from its differences, especially the least advantaged, then that society is becoming more just over time. Also, if opportunities for personal or group mobility continue to remain open amidst the persistence of differences, then the overall sense of fairness and stability will tend to increase.

It is at this point that we can link the evolution of the brain to the emergence and ongoing expansion of both the theory and practice of justice. Cooperation, broad self-interest, generosity, concern for the well-being of others outside one's own self or group, consensus building, and resolving problems nonviolently are traits we associate with the cerebral cortex and not the autonomic nervous system. To this list we can add the development of the theory of justice, which embodies the characteristics that we associate with the newer part of the brain.

While the opposite tendency toward selfishness and the desire to control others also continues as an extension of the older part of the brain, during the process of evolution, these became balanced against, surpassed by, and even channeled through the predispositions that surfaced with the development of the newer part of the brain. While many examples can be cited to illustrate this point, we have shown in chapter 6 that one in particular stands out. In 1840, only two democratic countries that comprised 3.9 percent of the world's population existed. By the year 2000, this percentage had increased to 57.1 percent. While the eventual outcome of political changes occurring in the Middle East, called the Arab Spring that began in 2011, and elsewhere throughout the world is unknown as of this writing, what is clear is that the desire for democracy among an increasing percentage of the world's population continues its long term trend.

Thus, even though many cultural differences persist throughout the world, including older brain type behaviors that still drive the desire for conquest and domination in some people, the tendencies we associate with the newer part of the brain continue to grow. Why should this be so? The answer seems clear: all humans have brains, and because all brains include a cerebral cortex, the quest for justice as evidenced in the spread of democracy as well as other social improvements that extend from a foundation of scientific and technical knowledge is universal. As a result of this growing global trend, we conclude this section by setting our confidence level that the world is evolving under the guidance of and intelligent designer called God at 75 percent.

Humanity and Universal Morality

Building on the preceding summary of justice, in this section we turn our attention to the broader issue of the universal morality that is present in all societies and of which the theme of justice is an essential element. This area is of particular significance because of the role that the modern scientific research in the field of Anthropology has played in demonstrating that universal norms sit at the deepest level of all cultures as described in chapter 7. This is so despite the many pluralistic differences that catch the eye at first glance, such as dissimilar food choices, clothing tastes, and a host of other specific preferences.

During the past 150 years, the discipline of Anthropology has progressed through two distinct phases. Using their modern method of field research, late nineteenth and early twentieth century anthropologists fanned out into many of the non-Western small and isolated pre-modern societies. These anthropologists were the first to record the range of differences that exist from culture to culture. As a result of their discoveries, the phrase cultural relativism came to describe the cross cultural variations that they observed. The works of Ruth Benedict and Margaret Mead epitomized this early phase.

Based on the studies of this early generation of anthropologists, it was but a short step to go from the science of cultural relativism to the philosophy of moral relativism, which presupposes that universal values do not exist. Instead, many concluded that values are merely relative to the individuals or societies that espouse them. This implies that cross-cultural judgments of any given group's normative preferences are invalid because no universal ethical framework is available to evaluate them. Kindness and killing stand on equal moral footing, and no objective standards can be used to condemn practices such as Nazi genocide against the Jews, slavery wherever it appears on the planet, or any other form of cruelty.

As we pointed out in chapter 7, it is illogical to advocate tolerance of cultural differences while being intolerant of cultures or individuals who express intolerance or hatred of others. In addition to describing the political context from which this contradiction arose, we demonstrated that it was through the ongoing field research of the later generation of anthropologists during the post-colonial era that the scale began to tip away from moral relativism and toward moral universalism. This shift occurred as a result of revisiting formerly investigated cultures, as in the case of those explored by Margaret Mead, as well as examining unstudied ones.

As newer research began to emerge, it revealed that despite the diverse cultural practices that vary across societies, they all appeared to possess a deeper set of shared values. Further studies took this discovery one step farther. They began to demonstrate that at the deepest level among all the cultures there is a single unifying ethical principle called the Universal Reciprocity Norm that gives rise to all other shared standards including the universal quest for justice that we discussed in chapter 6 and in the previous section of this chapter. Unlike moral relativism, this Universal Norm also serves as the foundation for differentiating kindness from killing and for condemning genocidal behavior and slavery anywhere on Earth.

The evidence, both old and new, for the existence of this norm is widespread. In ancient times, it took the form of the golden version (Do to others as you want them to do to you.) and its alternative expression the silver version (Do not do to others what you do not want them to do to you.). Both of these versions of the Universal Reciprocity Norm emerged thousands of years ago among isolated ancient cultures that had no contact or knowledge of each other's existence. The most important modern philosopher Immanuel Kant called this norm the Categorical Imperative that he was confident applied to everyone. The United Nations' Declarations of Universal Rights and the Parliament of the World's Religions' statements on Global Ethics are not only extensions of this Norm, but they embody many other universal moral principles as well.

Thus, the evidence is strong. Indeed, it is persuasive that a ubiquitous Reciprocity Norm exists and runs like a thread that ties together the pluralistic cultures of the world—from ancient to modern. See Table 11 in chapter 7. At the same time, continuing hostilities around the Earth indicate that we have not yet entered the front door of moral perfection and related behavior. The reason for this seems clear: Despite the evolution of the human brain, it still has two parts. The older part pushes us in the direction of narrow self-interest, while the newer part pulls us toward a broad self-interest that includes consequences that improve the well-being of others.

In light of this tension, where do we stand on the question of belief in the existence of God? At one level it confirms our earlier observation in chapter 1 that the universe is theologically ambiguous. Our behavior on Earth continues to be mix of both narrow selfishness and broad generosity. At the same time, given the compelling evidence that we humans have been aspiring for thousands of years to live by the Universal Reciprocity Norm that we have expressed in both its golden and silver formulations, despite our ongoing narrow self-centered behaviors, we are confident at the 80 percent level that an intelligent designer called God created and sustains the universe and is guiding human evolution on Earth.

L-M Scale Summary Table

Based on the above conclusions, we have included all of the L-M scale percentages in the summary Table 13.

As is readily apparent, Table 13 includes the six substantive areas that are the focus of the book: knowledge; the evolution of conscious, self-aware life; how non-life became life; the relationship of the brain to spirituality; the theme of justice; and the existence of universal morality.

Next to each of these six areas we have included a separate L-M scale along with the specific percentage, as specified in previous sections of this chapter, that refers to our level of confidence that this area contributes to the belief that an intelligent designer called God created the universe through the Big Bang and guided its development through a process of evolution that resulted in the emergence of conscious, self-aware life on Earth. We set our percentages after examining carefully a broad spectrum of knowledge that applies to each area. As Table 13 shows, we did not assign the same confidence level to each of the six areas. The levels vary by a margin of 35 percent with the lowest level being knowledge (60 percent) and the highest conscious, self-aware life (95 percent).

Table 13: The L-M scale.

Conclusion summary	L-M scale
Humanity is Structured for Knowledge	L ─── M ── 60%
The Universe Is Structured for Conscious, Self-Aware Life	L ─── M ── 95%
From Non-Life to Life	L ─── M ── 90%
The Brain Is Structured for Spiritual Experiences	L ─── M ── 80%
Humanity Is Structured for Justice	L ─── M ── 75%
Humanity Is Structured for a Universal Morality	L ─── M ── 80%
Overall total	L ─── M ── 80%

While Table 13 reveals the range of individual confidence levels, it is the combination of all of them that creates the most significant impression. We added together all of the percentages and divided this total number by six in order to reach an overall level of confidence, which is 80 percent. Another way to say this is that based on scientific evidence, we are 80 percent confident that an intelligent designer called God created the cosmos and guided its evolution.

It is essential to keep in mind that we do not derive our overall level of confidence through any one of the six areas by itself. It is the sum of the parts that provides the highest level of assurance. In other words, our confidence became cumulative as we advanced from one area of knowledge to another. At the same time, we have not set any one of the specific areas at 100 percent for the following reason. Because we believe that the universe is theologically ambiguous, we have not sought to either prove or disprove beyond a shadow of doubt that God exists. Instead, we have examined the findings of science in order to create a foundation of knowledge on which to build our confidence level at 80 percent.

We recognize that it is unavoidable that others will set different confidence levels and that everyone's judgment will contain personal and subjective elements. Nonetheless, given that we have based our 80 percent on the accumulation and integration of scientific knowledge from a broad range of areas, we believe that our conclusion is not uninformed, irrational, or arbitrary. Instead, it is scientifically persuasive.

In addition to averaging the six areas of knowledge that appear in the L-M Scale summary in Table 13, in Table 14 we go one step further by summarizing how the Anthropic Principle that we have used throughout the entirety of this book applies to both the physical and social worlds. As the left side of Table 14 shows, we have identified six themes that relate to these two worlds. They start with "simplicity of basic elements" and end with "amazement and wonder."

If the concept of randomness proved to be a superior theory for explaining the nature and destiny of the universe, then we would expect to find few similarities between the physical and social worlds. However, as Table 14 shows, this is not the case. Parallel patterns exist in abundance across both worlds. Furthermore, when we combine the results of Table 14 with the 80 percent confidence score that we derived by averaging the six individual scores in Table 13, our conclusion is further reinforced: it is far more probable that the origin and evolution of the universe is the handiwork of an intelligent designer called God than random forces.

Table 14: Anthropic Principle applied to the physical and human worlds.

	Theme	Physical world	Human world
1	Simplicity of basic elements	Universe is made of same basic material. The current Standard Model for sub-atomic particles and fields is widely accepted, appears to explain most of what is observed. See chapter 3.	A universal code of morality for the diversity of world cultures. Human quest for justice that cuts across all societies irrespective of their cultural or religious differences. Purposeful knowledge seekers who labor to learn as much as possible about how the world works.
2	Origins are simple, single sourced, and occur at one time.	One ultimate origin: Big Bang. Not multiple independent creation-type events spread out over time and space.	Widely accepted explanation is that anatomically modern *Homo sapiens* appeared in Africa 200,000 years ago.
3	Comprehensible	Mathematics, four forces, Standard Model, and other examples appear to explain most of the known universe's behavior.	Cooperation, abstract thought & language, spirituality & religion, knowledge, and populating the whole globe. Older & newer parts of the brain.
4	Evidence for structuring?	Yes, Anthropic Principle and associated discussion.	Yes, knowledge seeking, spirituality, justice, and morality.
5	Compatible with evolution?	Yes, development of elements in stellar interiors, development of universe, of life, and of the brain.	Yes, starting with basic survival skills 200,000 years ago, compared to where we are today.
6	An example for amazement and wonder?	Yes. Incomprehensible number of galaxies in all directions as detected by the Hubble Space Telescope.	Yes. From simple lifeless molecules 3.85 billion years ago to self-aware people who seek to understand the role of humanity in the universe.

Stage 3: Science and Belief in God

Next, based on our 1) presentation of evidence throughout the entirety of this book, 2) confidence levels as summarized in the L-M scale, and 3) application of the Anthropic Principle to both the physical sciences and human experience, we believe that we are in the midst of an evolving stage 3 regarding the relationship between the evolution of science and belief in

God. *We cannot overstate the importance of this point.* Stage 1 existed prior to the rise of modern science. During this stage, the world's many cultures and religious groups developed their diverse views of the origin and development of the universe and of humanity's role in it.

Stage 2 started with the rise of modern science and involved the reactions of many traditionalists to new knowledge that they perceived threatened their pre-scientific worldviews. As described in chapter 1, the writing of *The Fundamentals* in 1915 and the Scopes Monkey Trial in 1925 epitomized the conflict that crystallized into a standoff between the proponents of evolution and the advocates of biblical creationism and literalism. While many non-fundamentalists accepted some form of evolutionary theory, no matter where they stood on the conservative to liberal theological continuum, all theists rejected the belief that in order to practice science by using a naturalistic methodology that focuses on temporal and not formal, first, or final causes, it was necessary to accept a non-theistic explanation of the cosmos.

This being said, regardless of where theists of different stripes stood on the issues, more than any single event, it was the Scopes trial that dominated the headlines and captured the public's attention. The drama of this event obscured the deeper theist versus naturalist debate and turned the science and religion dispute into a standoff between biblical literalists and evolutionary modernists. The unfortunate legacy of this bitter clash was that in the popular mind one had to either accept science and reject God or accept God and reject science despite alternative views that existed at the time.

Many of the arguments that still exist between atheists and non-atheists stem from the ongoing reaction against the discoveries of modern science and in some cases even debates over definitions of the scientific method and how to apply it to various fields of inquiry. Our view is that these narrow stage 2 debates have reached a stalemate. They are backward looking and do not move us forward in a long overdue third stage that builds on recent scientific developments.

Stage 3 of the relationship between science and belief in God has been evolving during the past several decades. It has been progressive, which means that it could not have arrived at its present level until science had evolved to a more advanced stage where the evidence increasingly indicates that belief in God and the practice of science are not opposed to each other. As we have made clear throughout this book, the best evidence available enhances our confidence that recent scientific knowledge supports belief in God.

Evolution and Belief in God

Our next observation dovetails with our belief that the best available scientific knowledge, and not the lack of it due to gaps and complexities that are inherent roadblocks, reinforce belief in God. Whereas many religious conservatives rejected and continue to reject the theory of evolution, stage 3 incorporates it completely and combines it with theism. As is clear, the best available research discoveries that we have included in our discussion build on and are fully integrated into the patterns of cosmic and earthly evolution.

As we indicated earlier, there is ample evidence to support belief in evolution. Despite unresolved questions related to how the physiology of the brain connects to thinking, knowledge, and spirituality, we have presented evidence for the brain's evolution in chapter 5. On the basis of knowledge accumulated through scientific research, we see that the human brain has evolved into two distinct spheres. The first is the older, lower, or reptilian part called the autonomic nervous system; and the second, which evolved later, is the newer, higher, and more advanced segment called the cerebral cortex.

Each domain of the brain is associated with different emotional and behavioral tendencies. The reptilian area is tied to narrow and exclusive self-interest or selfishness, competition, conflict, individual and group domination through winning or conquest, and an inclination to settle differences through violence if necessary. In its most primordial forms of expression, it has served to protect and continues to foster human survival in the face of threatening hostilities.

At the same time, as a result of the brain's evolution, the cerebral cortex has given rise to other tendencies. These include generosity, broad self-interest that includes considering the needs of others outside of one's own self or group, cooperation, survival by building consensus, and a preference for resolving conflict through nonviolence. In addition, the newer sphere of the brain is essential for developing the capacities for spirituality, justice, and universal morality that appear in all societies. It also facilitates the accumulation and retention of the knowledge foundation through which we comprehend the origins and operations of our universe and which we use for ongoing adaptation and advancement as a species.

When all of the best pieces of available evidence are bundled together, we are left with one overarching conclusion: accepting evolution in all its forms—from using it to describe the development of the universe from the moment of the Big Bang to the Darwinian view of life on Earth—is not a threat to belief in God. Furthermore, belief in evolution does not obligate one to adopt a non-theistic philosophy of life. Nor does engaging in scientific research necessitate a rejection of belief in God. The contrary is true. Belief in God, the practice of science, and the acceptance of evolution are mutually reinforcing as demonstrated by the advancement of scientific knowledge itself.

Conclusion

Our conclusion is brief and to the point. As we have indicated repeatedly, we believe that the universe is theologically ambiguous with regard to either proving or disproving the existence of God. Pro and con arguments have existed for centuries with no definitive resolution one way or the other. As a result, we have chosen a different pathway that entails setting confidence levels based on the best scientific evidence available in various field of inquiry. After applying the Anthropic Principle to similarities that exist between the physical and human worlds in Table 14 and setting confidence levels for the L-M scale in Table 13, it has become clear to us that we have entered a new stage 3 in which scientific knowledge and belief in God are mutually supportive and where evolution plays a central role.

Finally, building on the multiple areas of knowledge that are included in this book, we believe that explaining the origin and destiny of the universe in terms of randomness is far less compelling than explaining it in terms of an intelligent designer called God who created the cosmos and guided its evolution toward conscious, self-aware life on Earth. It is for this reason that we conclude at the 80 percent confidence level that belief in God is here to stay.

Epilogue

This Epilogue focuses on issues that call for follow up discussion. In the process of examining the relationship between scientific knowledge and belief in God, we have written very little about the nature of God and God's relationship to humanity beyond identifying God as the intelligent designer who created the cosmos and directed its development. We have said nothing about the various views of life after death. Nor have we reexamined ancient texts or argued against the philosophical assumptions of naturalism as alternative pathways to exploring the question of God's existence. We have also steered clear of identifying our view of God with any of the theological concept that are associated with the world's religions.

At the same time, we want to avoid the implication that our view of an intelligent designer who created the universe is the same as the conceptions of God that are associated with the theology of Deism or with Aristotle's Unmoved Mover. In both of these cases, God is perceived as a kind of cosmic first cause watch maker who wound up the world—so to speak—to get it started and then stood back as it evolved through its own natural laws. In the Deist view God's role is limited to being the creator of the physical universe with no follow up intervention. For Aristotle, believing in an Unmoved Mover who started the universe is a logical necessity even though it is not necessarily an empirical one.

Our approach to belief in God does not focus on philosophical assumptions or theistic deductions but instead by drawing inferences after examining scientific evidence. It is from this empirical platform that we arrive at our conclusion about God's nature as an intelligent designer who created the universe through the Big Bang. In addition to being the creator, this intelligent designer also sustained the universe and guided its development through the process of evolution that led to the emergence of conscious, self-aware life on Earth.

We recognize that the new atheists and other non-theists champion a naturalistic interpretation based on the notion of randomness. However, we have a high degree of confidence that our case for the existence of a creator—designer God is more persuasive than theirs. We base our position and confidence levels entirely on scientific evidence, not on the lack of it. This means that we rooted our conclusion in inductive reasoning and observations that include the narrow allowable ranges for the mathematical constants by which the physical universe functions, the development of the new part of the brain that fosters generosity and cooperation, the quest for global justice, and the pervasive presence of the Universal Reciprocity Norm.

At the same time, we are not in a position in this book to go beyond this point. In particular, we do imply that our conclusion about the nature of God and God's involvement in the creation and evolution of the universe supports any given religion's view of God. It might or it might not. Just as we have made our case for the relationship between science and belief in God, others might want to take the next step and show how this position ties to their theological traditions—as many already have and will continue to do. We strongly encourage others who are associated with the diverse faith traditions throughout the world, such as Christianity, Judaism, Islam, Hinduism, Buddhism, among others, to take or continue taking this next step.

Bibliography

Alencar, Silvia H. P., et al. "The Spectral Variability of the Classical T Tauri Star DR Tauri." *Astronomical Journal* 122 (2001) 3335–60.

Anderson, Don L. "The Earth as a Planet: Paradigms and Paradoxes." *Science* 223 (1984) 347–55.

Aristotle, "Nicomachean Ethics," in *The Basic Works of Aristotle*, edited by Richard McKeon. New York: Random House, 1941.

Barbour, Ian. *Religion and Science: Historical and Contemporary Issues*. San Francisco: Harper, 1997.

———. *When Science Meets Religion*. San Francisco: Harper, 2000.

Barrow, John D., and Frank J. Tipler. *The Anthropic Cosmological Principle*. New York: Oxford University Press, 1986.

Bell, Wendell. *Foundations of Futures Studies: Human Science for a New Era* .Volume 2. New Brunswick, NJ: Transaction, 1997.

Benedict, Ruth. *Patterns of Culture*. Boston: Houghton Mifflin, 1953.

Blakstad, Oskar. "Research Methodology: Key Concepts of the Scientific Method." No pages. Online: http://www.explorable.com.

Boaz, Franz. *The Mind of Primitive Man*. New York: Collier Books, 1963.

Brown, Donald E. *Human Universals*. Philadelphia: Temple University Press, 1991.

Campbell, I. H., and S. R. Taylor. "No Water, No Granite—No Oceans, No Continents." In *Geophysical Research Letters* 10 (1983) 1061–64.

Carter, Brandon. "The Anthropic Principle and Its Implications for Biological Evolution." In *Philosophical Transactions of* the *Royal Society of London, Series A* 310 (1983) 352–63.

Chomsky, Noam. *Language and Mind*. Enlarged Edition. New York: Harcourt, Brace, Jovanovich, 1972.

Clayton, Philip. *The Oxford Handbook of Religion and Science*. New York: Oxford University Press, 2008.

Collins, Francis S. *The Language of God: a Scientist Presents Evidence for Belief*. New York: Free Press, 2006.

———. "Making Life." Online: http://www.pbs.org/wgbh/nova/tech/collins-genome .html.

Cottrell, Ron. *The Remarkable Spaceship Earth*. Denver: Accent, 1982.

Cox, Brian, and Andrew Cohen, *Wonders of the Universe*. San Francisco: HarperCollins, 2011.

Dampier, William Cecil. *A History of Science and Its Relationship to Philosophy and Religion*. 4th ed. Cambridge: Cambridge University Press, 1979.

Darwin, Charles. *Origin of Species*. New York: D. Appleton, 1859.

———. *The Descent of Man*. New York: D. Appleton, 1880.

David R. Soderblom, et al. "Rotational Studies of Late-Type Stars. VII. M34 (NGC 1039) and the Evolution of Angular Momentum and Activity in Young Solar-Type Stars," *Astrophysical Journal* 563 (2001) 334–40.

Dawkins, Richard. *The God Delusion.* New York: Bantam, 2006.

————. *The Greatest Show on Earth.* New York: Free Press, 2009.

De Gray, Aubrey, and Michael Rae. *Ending Aging: The Rejuvenation Breakthroughs That Could Reverse Human Aging in Our Lifetime.* New York: St. Martin's, 2007.

Diamond, Stanley. *In Search of the Primitive.* New Brunswick, N.J.: Transaction, 2004

Durkheim, Emile. *The Elementary Forms of the Religious Life.* Translated by Karen E. Fields. New York: The Free Press, 1995.

Emmett, Arielle. "The Human Genome." In *The Scientist: Exploring Life, Inspiring Innovation.* June 24, 2000.

"Explorable: Scientific Method: How Knowledge Is Made." No pages. Online: http://www.explorable.com.

Frostig, R. D. "Functional Organization and Plasticity in the Adult Rat Barrel Cortex: Moving out-of-the-box." *Current Opinion Neurobiology* 16 (2006) 1–6.

Gaiarsa, J. L., et al. "Long-term plasticity at GAB Aergic and Glycinergis Synapses: Mechanisms and Functional Significance." *Trends Neuroscience* 25 (2011) 564–70.

Gale, George. "The Anthropic Principle." *Scientific American* 245 No.6 (1981) 154–71.

Goddard, Jolyon Goddard, editor. *Concise History of Science & Invention: an Illustrated Time Line.* Washington, D.C.: National Geographic, 2010.

Graham, Sarah. "Clays Could Have Encouraged the First Cells to Form." *Scientific American* (October 24, 2001). Online: http://www.sciam.com/article.cfm? Id=clay-could-haveencourage.

Granoff, Jonathan. "Peace and Security." In *Analyzing Moral Issues,* edited by Judith A. Boss. 5th edition. 627–30. New York: McGraw-Hill, 2010.

Greenstein, George. *The Symbiotic Universe: Life and Mind in the Cosmos.* New York: William Morrow, 1988.

Gribbin John. *Alone in the Universe: Why Our Planet Is Unique.* Hoboken, NJ: John Wiley & Sons, 2011.

Gubbin, John. "The Origin of Life: Earth's Lucky Break." *Science Digest* May (1983) 36–102.

Hagerty, Barbara Bradley, *Fingerprints of God: The Search for the Science of Spirituality.* New York: Riverhead, 2009.

Hamer, Dean. *The God Gene: How Faith is Hardwired into Our Genes.* New York: Doubleday, 2004.

Hammond, Allen H. "The Uniqueness of the Earth's Climate" *Science* 187 (1975) 245.

Hanisch H. "Children's and Young People's Drawings of God (lecture given at the University of Gloucestershire.)" Online: http://www.uni-leipzig.de/~rp/vortraege/hanisch01.html.

————. "*The graphic development of the God picture with children and young people: An empirical comparative investigation with religious and non-religiously educating at the ages of 7–16.*" Stuttgart and Leipzig: University of Leipzig, 1996.

Harms, E., "The development of religious experience in children." *American Journal of Sociology.* 50 (1944) 112–22.

Harris, Sam. *The End of Faith: Religion, Terror, and the Future of Reason.* New York: W. W. Norton, 2004.

Hart, Michael H. "Habitable Zones about Main Sequence Stars." *Icarus.* 37 (1979). 351–357.

———. "The Evolution of the Atmosphere of the Earth." *Icarus* 33 (1978) 23–39.

Herbst, Judith. *A History of Transportation: Major Inventions through History.* Minneapolis: Twenty-First Century Books, 2006.

Hermanns, William. *Einstein and the Poet: in Search of the Cosmic Man.* Brookline Village, MA: Branden, 1983.

Hitchens, Christopher. *God Is Not Great: How Religion Poisons Everything.* New York: Twelve, 2007.

Horgan, John. *The End of Science: Facing the Limits of Knowledge in the Twilight of the Scientific Age.* New York: Broadway, 1997.

Hoskin, Michael, editor. *The Cambridge Concise History of Astronomy.* Cambridge: Cambridge University Press, 1999.

"How Did Life Begin?" Online: http://www.pbs.org/wgbh/nova/evolution/how-did-life-begin.html.

How Life Began. A&E Television Networks, LLC, 2008.

Howland, John. *The Surprising Archaea: Discovering Another Domain of Life.* New York: Oxford University Press, 2000Howland, John. "Vents and Volcanoes." Online: http://www.oceantoday.noaa.gov/underwatervolcnoes/welcome.html.

"Hubble Goes to the eXtreame to Assemble Farthest-Ever View of the Universe." Online: http://www.nasa.gov/mission_pages/hubble/science/df.html.

"Hundreds of Proofs of God's Existence (Formerly: Over Three Hundred Proofs of God's Existence)." Online: http://www.godlessgeeks.com/LINKS/GodProof.htm.

Hurtley, Stella M., and Elizabeth Pennisi. "Journey to the Center of the Cell." *Science* 318 (2007) 1399.

Kant, Immanuel. *The Critique of Practical Reason.* Translated by Lewis White. 1992, Upper Saddle River, NJ: Prentice-Hall, 1992.

———. *Critique of Pure Reason.* Translated by F. Max Muller. New York: Anchor, 1966.

———. *Fundamental Principles of the Metaphysics of Ethics.* Translated by Thomas Kingsmill Abbott. London: Longmans, Green and Co., Ltd., 1926.

Keith, Kent M. *Morality and Morale: A Business Tale.* Honolulu: Terrace Press, 2012.

Kidder, Rushmore M. *Shared Values for a Troubled World.* San Francisco: Jossey-Bass, 1994.

Kluckholn, Clyde. *Mirror for Man.* New York: Fawcett, 1944.

Kroeber, Alfred. A*nthropology.* New York: Harcourt and Brace, 1949.

Kluger, Jeffrey. "Is God in Our Genes?" *Time Magazine.* October 25 (2004) 62–64.

Knoll, Andrew. *Life on a Young Planet: The First Three Billion Years of Evolution on Earth.* Princeton University Press, 2003.

Kreeft, Peter, and Ronald K. Tacelli. "Twenty Arguments for the Existence of God." Online: http://peterkreeft.com/topics-more/20_arguments-gods-existence.htm.

"Large hadron Collider." Online: http://www.public.web.cern.ch/public/en/lhc/lhc-en.html.

Lee, Harper. *To Kill a Mockingbird.* Philadelphia: J.B. Lippincott, 1960.

Leeming, David Adam, and Margaret Adams Leeming. *Encyclopedia of Creation Myths.* 2nd ed. Santa Barbara, CA: ABC-CLIO, 1994.

———. *A Dictionary of Creation Myths.* New York: Oxford University Press, 2009.

Lewis, C. S. *Mere Christianity.* A revised and enlarged edition. New York: Collier, 1952.

Bibliography

"Major Religions of the World Ranked by Number of Adherents." Online: http://www. adherents.com/Religions By Adherents.

Manning, Craig E. et al. "Geology and Age of Supracrustal Rocks, Akilia Island, Greenland: New Evidence for a >3.83 Ga Origin of Life." *Astrobiology* 1 (2001) 402–3.

McFaul, Thomas R. *The Future of God in the Global Village: Spirituality in an Age of Terrorism and Beyond.* Bloomington, IN: AuthorHouse, 2011.

———. *The Future of Peace and Justice in the Global Village: The Role of the World Religions in the Twenty-First Century.* Westport, Conn: Praeger, 2006.

———. *The Future of Truth & Freedom in the Global Village: Modernism and the Challenges of the Twenty-First Century.* Santa Barbara, CA: Praeger, 2010.

Mead, Margaret. *Coming of Age in Samoa.* New York: William Morrow, 1928.

———. *Sex and Temperament.* New York: William Morrow, 1935.

"Meteorite That Fell in 1969 Still Revealing Secrets of the Early Solar System." Online: http://www.scientificamerican.com/article.cfm?id=murchison-meteorite.

Munson, Ronald. *Intervention and Reflection: Basic Issues in Bioethics.* 9th ed. Boston: Wadsworth, 2012.

"NASA Researchers Make First Discovery of Life's Building Block in Comet." Online: http://www.nasa.gov/mission_pages/stardust/news/stardust_amino_acid.html.

Nelson, K. *The Spiritual Doorway in the Brain: A Neurologist's Search for the God Experience.* New York: Penguin, 2011.

Newberg, A, and M. R. Waldman. *How God Changes Your Brain.* New York: Ballantine, 2009.

Nisbet, Robert. *History of the Idea of Progress,* Piscataway, NJ: Transaction, 1980.

Onfray, Michel. *Atheist Manifesto: the Case against Christianity, Judaism, and Islam.* New York: Arcade, 2007.

Owen, Tobias et al. "Enhanced CO_2 Greenhouse to Compensate for Reduced Solar Luminosity on Early Earth." *Nature* 277 (1979)640–41.

Parliament of the World's Religions, *Toward a Global Ethic: an Initial Declaration.* New York: 1993.

Peterson, Michael, et al. *Philosophy of Religion: Selected Readings.* New York: Oxford University Press, 2007.

Poe, Marshall T. *A History of Communications: Media and Society from the Evolution of Speech to the Internet.* Cambridge: Cambridge University Press, 2011.

Polkinghorne, John. *Theology in the Context of Science.* New Haven: Yale University Press, 2009.

Polley, D. B., et al. "Naturalistic Experience Transforms Sensory Maps in the Adult Cortex of Caged Animals" *Nature* 429 (2004) 67–71.

Porter, Roy. *Blood and Guts: A Short History of Medicine.* New York: W. W. Norton, 2004.

"Quantum Universe: The Revolution in the 21st Century." Online: http://www.interactions .org/cms/?pid=1012346.

Quoted in Barbara Bradley Hagerty, *Fingerprints of God: the Search for the Science of Spirituality.* New York: Riverhead, 2009.

Rana, Fazale, and Hugh Ross. *Origins of Life.* Colorado Springs: NavPress, 2004.

Rawls, John. *A Theory of Justice.* Cambridge: Harvard University Press, 1971.

Rhode, Katherine L., et al. "Rotational Velocities and Radii of Pre-Main-Sequence Stars in the Orian Nebula." *Astronomical Journal* 122 (2001) 3258–79.

Ricardo, Alonso, and Jack W. Szostak. "Life on Earth," *Scientific American.*September(2009) 54–61.

Rood, Robert T., and James S. Treffi. *Are We Alone? The Possibility of Extraterrestrial Civilizations*. New York: Charles Scribner's Sons, 1983.

Ross, Hugh. *Genesis One: A Scientific Perspective*. Pasadena, CA: Reasons to Believe, 1983.

Rovner, Sophie L. "Bacteria Boast Unexpected Order (staff written summary)."*Chemical and Engineering News*. Februrary 16 (2009) 42–46.

Ruthvan, Malise. *Fundamentalism: the Search for Meaning*. New York: Oxford University Press, 2004.

Segal, Nancy L. *Born Together—Reared Apart: The Landmark Minnesota Twin Study*. Cambridge, MA: Harvard University, 2012.

Shermer, Michael. *The Believing Brain*. New York: Times Books, 2011.

Soderblom, David, et al. "Rotational Studies of Late-Type Stars. VII. M34 (NGC 1039) and the Evolution of Angular Momentum and Activity in Young Solar-Type Stars." *Astrophysical Journal* 563 (2002) 334–40.

Solomon, Robert C. *Introducing Philosophy: a Text with Integrated Readings*, edited by Robert C. Solomon. 8th edition. New York: Oxford University Press, 2005.

Stannard, Russell. *Science & Belief: The Big Issues*. Oxford, UK: Lion Hudson, 2012.

Stark, Rodney. *The Triumph of Christianity*. New York: HarperCollins, 2011.

Stenger, Victor J. *God: the Failed Hypothesis—How Science Shows That God Does Not Exist*. Amherst, NY: Prometheus, 2006.

Templeton, John M. "God Reveals Himself in the Astronomical and in the Infinitesimal." *Journal of the American Scientific Affiliation* December (1984) 196–98.

Ter Harr, D. "On the Origin of the Solar System." *Annual Review of Astronomy and Astrophysics*. 5 (1985) 267–78.

The Large Hadron Collider. Online: http://public.web.cern.ch/public/en/lhc/lhc-en.html.

Toon, Owen B., and Steve Olson. "The Warm Earth." *Science* 85 (October1985) 50–57.

van Doren, Charles. *A History of Knowledge: Past, Present, and Future*. New York: Random House, 1991.

Velasquez, Manuel G. *Business Ethics: Concepts and Cases*. 7th ed. Upper Saddle River, NJ: Pearson, 2012.

Wait, Marianne. "No More Brain Drain: Proven Ways to Maintain Your Mind & Memories." In *Reader's Digest*, 2009.

Walter, Frederick M., and Don C. Barry. "Pre-and Main-Sequence Evolution of Solar Activity." In *The Sun in Time*, edited by C. P. Sonett et al., 633–57. Tucson: University of Arizona Press, 1991.

Walzer, Michael. *Spheres of Justice: a Defense of Pluralism and Equality*. New York: Basic Books, 1983.

Ward, William R. "Comments on the Long-Term Stability of the Earth's Obliquity." *Icarus* 50 (1982) 444–48.

Wells, Llyd E. et al. "Reseeding of Early Earth by Impacts of Returning Ejecta During the Late Heavy Bombardment." *Icarus*. 162 (2003) 38–46.

"What People Do and Do Not Believe In." Harris Poll. No pages. Online: http://www .harrisinteractive.com/vault/Harris_Poll_2009_12_15.pdf.

Index